T0139862

Studies in Computational Intelligence 478

Editor-in-Chief

For further volumes:
http://www.springer.com/series/7092

Maciej Krawczak

Multilayer Neural Networks

A Generalized Net Perspective

Prof. Maciej Krawczak
Polish Academy of Sciences
Warsaw School of Information Technology
Systems Research Institute
Warsaw
Poland

ISSN 1860-949X ISSN 1860-9503 (electronic)
ISBN 978-3-319-03390-7 ISBN 978-3-319-00248-4 (eBook)
DOI 10.1007/978-3-319-00248-4
Springer Cham Heidelberg New York Dordrecht London

Softcover re-print of the Hardcover 1st edition 2013

Printed on acid-free paper

Springer is part of Springer Science+Business Media (www.springer.com)

Foreword by Krassimir T. Atanassov

The purpose of this book is threefold. Firstly, at the beginning, to show that multilayer neural networks can be considered as multistage systems, and, as a result, the learning algorithms can be treated as particular multistage optimal control problems. Learning problems are characterized by high dimensionality, and the optimal control theory methodology must be modified and adopted for this particular optimisation problem.

The heuristic dynamic programming algorithm, one of the learning algorithms proposed by the author, provides a near optimal solution of the performance index of a learning error. It seems that the algorithm is able to avoid local minima that often appear during the learning process.

Also, the adjoint neural networks are defined and they give direct expressions for gradient-based learning algorithms, which could be implemented as electronic devices.

Secondly, using the simulation process and learning process, the author developed generalized net descriptions of the processes. This way, the author has proved that the concept of a generalized net, developed by myself for the last 30 years, can be successfully used for the modelling of real systems. The concept of a generalized net has been proposed, in an abstract formalism, as unification of various modifications of the Petri nets.

Multilayer neural networks are well known tools for deriving many complex nonlinear models, since they are commonly used as universal approximators and universal pattern classifiers. Therefore, displaying generalized net methodology through developing new descriptions of the neural networks processes seems to be a very good idea. The book contains several parts with detailed generalized net descriptions of such processes as the simulation and learning for various algorithms. It has been shown that the formal abstract description of the generalized nets may be used for the description of logic organization of the multilayer neural networks as well as a uniform representation of the learning algorithms. The developed description of all formal elements of the generalized net models makes it possible to better understand the very essence and functioning of multilayer neural networks.

Thirdly, and what is the most important for me, the book gives the answer to the open question, which I formulated in November 1991: whether the generalized nets can be used as an universal method for modelling different artificial intelligence tools, among the functioning of all neural networks.

I have known Maciej Krawczak for many years. He is one of the colleagues, who are the most (in the world) involved in research regarding the generalized net theory and applications.

Professor Krassimir Atanassov
D.Sc., D.Sc.
Academician
Bulgarian Academy of Sciences

Preface

In the rich literature dealing with artificial neural networks, little attention has been paid to the consideration of neural networks from the system theory point of view.

The first conception of a system with many simple processing elements and a hypothetical algorithm for learning was given by Pitts and McCulloch (1947), and the present-day supervised learning paradigms are based on this idea.

The delta rule for learning, introduced by Widrow and Hoff (1960), leading to the iterative procedures for connection weights adjustment was fundamental for the major learning algorithms.

Now, considering the backpropagation algorithm, we would like to mention (except for Werbos (1989), and Rumelhart, Hinton and Williams (1986)) that the pioneering works of e.g. Kelley (1960), and the supporting sequel of Dreyfus (1990), were very similar to the backpropagation but obtained for the optimal control theory problems.

Krawczak and Mizukami (1994) wrote the very first paper about consideration of multilayer neural networks as multistage systems.

This approach was then developed by the author of this book in several scientific works (Krawczak 1995a, 1997, 1998, 1999a, 1999b, 2000a, 2001a, 2001b, 2002b, 2002c, 2003a, 2003b, 2003e). Additionally, in these works the author developed several learning algorithms formulated as the dynamic programming as well as multiobjective optimisation problems.

Some works by Atanassov (1997, 1998), Atanassov and Aladjov (2000), and Krawczak, Aladjov and Atanassov (2003a, 2003b), as well as by Krawczak (2003a, 2003b, 2003e) are devoted to the development of the generalized net description of neural networks simulation process and neural networks learning process.

The primary purpose of this book is to show that a multilayer neural network can be considered as a multistage system, and then that the learning of this class of neural networks can be treated as a special sort of the optimal control problem. In this way, the optimal control problem methodology, like dynamic programming, with modifications, can yield a new class of learning algorithms for multilayer neural networks. The gist of the new class of learning algorithms consists in aggregating neurons within separate layers, thus the layers become stages, while weights become controls. The problem of optimal weight adjustment is converted to a multistage optimal control problem. The developed *heuristic dynamic programming* algorithm allows for avoiding of local minima of the performance index of learning more often than the commonly used backpropagation algorithm (with modifications), while the development of *the adjoint neural networks* allows for finding expressions for gradient based learning algorithms directly.

Another purpose of this book is to show that the generalized net theory developed by Atanassov (1984) as the extension of the ordinary Petri net theory and its modifications can be successfully used as a new description of multilayer neural networks. Several generalized net descriptions of neural networks functioning processes are considered, namely: the simulation process of networks, a system of neural networks, the learning algorithms developed in this book, the backpropagation algorithm, and the adjoint neural networks. The use of the generalized nets methodology shows a new way to describe functioning of discrete dynamic systems. Consideration of neural networks dynamics is only an excuse to show functionality of the generalized net methodology.

The scope of this book is as follows: in Chap. 1, we briefly present the properties of the idea of sequential computation based on *the von Neumann machine* (von Neumann 1966). In the alternative idea of *the functional processing*, the data processing is performed in *parallel*, realized by e.g. *artificial neural networks*, especially the *multilayer* ones, characterized by a number of interconnected *neurons*, *weights* and *learning* processes or *training* from a *set of examples*. In a comprehensive form, the main consequences of multilayer neural networks are pointed out, such as *simulation*, *modelling*, *learning* and *generalization* problems. For convenience, a slightly different notation is introduced for describing the neuron connections.

In Chap. 2, a short introduction to generalized nets theory is recalled (Atanassov 1991, 1998), namely, the concept of generalized nets, the algorithm of generalized nets, the algebraic aspects of generalized nets, the operator aspects of generalized nets, and a short review of generalized nets topics.

In Chap. 3, the simulation process of multilayer neural networks is recalled and the generalized net description is developed. It is shown that the neural network simulation process can be described in terms of the generalized nets methodology.

In Chap. 4, we present problems of the process of learning from examples: set of $\{input, output\}$ pairs, *the performance indices* of learning, *the activation functions* of neurons, *the delta rule* (Widrow and Hoff 1960), *the generalized delta rule* (Rumelhart, Hinton and Williams 1986), *the backpropagation algorithm* that has roots in the early works related to optimal control theory algorithms (Kelley 1960, Dreyfus 1990, Krawczak 2000a, 2001b, Krawczak and Mizukami 1994). The chapter contains generalized net description of the backpropagation algorithm.

Chap. 5 is devoted to consideration of the learning of neural networks as *a multistage control problem* as well as application of *dynamic programming* methodology (Krawczak 1999a, 1999b, 2000a, 2001b). The problem of optimal weight adjustment is converted into a problem of optimal control. The return functions for each layer are defined, and minimization of these functions is performed layer by layer, starting from the last layer.

In Chap. 6, *a gain parameter* to the sigmoidal function neuron models is introduced. *The continuation methodology* is used in order to change the value of the gain parameter (Krawczak 1999b, 2000b, 2000c, 2000e). On the basis of the first order differential programming algorithm, the procedure to find the optimal value of the gain parameter is presented (Krawczak 2001a, 2002b). Due to the used methodology, *the heuristic dynamic programming algorithm* for neural network learning, as well as the generalized net description of the algorithm (Krawczak 2003a, Krawczak 2003b), is developed.

In Chap. 7, the learning process of neural networks is considered from the point of view of the graph theory. Neural networks treated *as flow graphs* give very interesting and new properties of the neural networks learning process. The approach is based on the Tellegen's theorem (Tellegen 1952) used in the electric circuits. The graph methodology incorporates *the reciprocal graph,* and using terminology adopted from the optimal control theory (Bryson and Ho 1969), such neural networks in which signals flow in the opposite direction are called *the adjoint neural networks* (Krawczak 2002a), and the generalized net of the adjoint neural networks is developed.

Finally, a concluding summary is presented (Chap. 8).

At the end, there is a rich relevant bibliography on the neural networks, optimisation theory, control theory as well as generalized net theory.

The list of people to whom I would like to express my gratitude is very long, let me therefore mention only a few of them here. I would like to thank Professor Janusz Kacprzyk for drawing my attention to the generalized net theory, and for soft and permanent encouragement. I am grateful to Professor Krassimir Atanassov for the initiation to the generalized net theory, for inspiration, stimulation, and fruitful ideas, as well as to his colleagues for fruitful collaboration.

I would like to express my gratefulness to Professor Zdzisław Pawlak who wrote an appreciative opinion of my first book related to generalized net methodology, it was done on the occasion of the Minister of Science and Higher Education Award of Poland, the opinion date is December 1, 2005.

Last but not the least, I am greatful to my family for their support and patience.

Warsaw,
December 2012

Maciej Krawczak

Contents

Chapter 1
Introduction to Multilayer Neural Networks

1.1 Sequential versus Parallel Computation

The notion of sequential computation is based on the von Neumann machine (von Neumann 1966) whose functioning can be displayed in a very simple, symbolic way (Fig. 1.1). The first part of the respective machinery, called memory, serves as a store of computer programmes with instructions, whereas the second part, called processor, in which an ordinary arithmetic rule of addition is performed within an operation system. The computer works in the following steps:

1. An instruction from the memory is taken to the processor.
2. Data required by the instruction is taken from the memory.
3. The instruction is performed.
4. Results are sent to the memory.
5. Go to step 1.

The computer works sequentially in two ways. The processor is fed the data instruction-by-instruction as well as piece-by-piece, and the results are stored in the memory. The processor also works sequentially.

Computer programs describe in a formal way the procedures for solving the problems considered. The basic feature of conventional computers is operating with numbers, exactly speaking with digits. Computation is sensitive to the accuracy of data as well as hardware reliability.

There is another way to process data, namely *functional processing*. It is not sequential but *parallel*. Instead of numbers a "new computer" can process symbols according to some rules describing a system. This is, perhaps, a way of imitation of functioning of a brain. One possible method to fill out conditions of massive or parallel computation is the methodology of *artificial neural networks*, especially *feedforward* and *multilayer* neural networks. This class of neural networks is characterized by a system of a number of basic elements called *neurons* organized in *layers*, which are interconnected. The neural networks processing or computation ability depends on the connection strengths of neurons, called *weights*, obtained during *learning* or *training* from a *set of examples*.

M. Krawczak: *Multilayer Neural Networks*, SCI 478, pp. 1–14.
DOI: 10.1007/978-3-319-00248-4_1

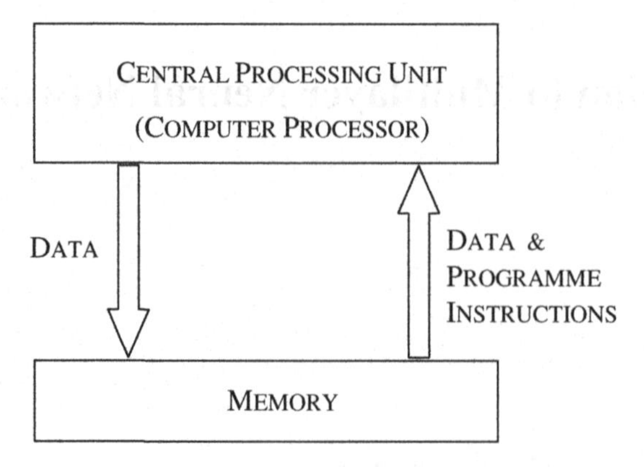

Fig. 1.1 Schematic idea of conventional computers (von Neumann machine)

In such neural networks, computation is distributed throughout a network so that it might be carried out nearly simultaneously. Hardware implementation of the considered architecture of neural networks asserts massively parallel data processing and in result high speed. Additionally, the high speed of data processing is assured by the simplicity of the neuron model. Another feature of the neural networks is *robustness* with respect to completeness and noisiness of data as well as hardware reliability. There are the well-known *generalization* properties of neural networks and low sensitivity to a neuron failure. After training of a network with many neurons, it is possible to remove some of them, and the result of computation will be little changed.

These features caused high interest in neural networks and their applications in pattern classification, function approximation, signal processing and control.

This chapter is devoted to some preliminaries of the neural networks with emphasis on the multilayer feedforward neural networks properties. The neural networks of this kind perform the parallelly distributed processing of information as a counterpart to the sequential information processing used in the conventional computers. We will use a little bit different notation than in the common literature of the subject, and when we use the notation of neural network we will mean the multilayer feedforward neural network (Krawczak 2003b).

1.2 Simulation Process of Neural Networks

A multilayer neural network, an example is shown in Fig. 1.2, consists of a number of elementary processing units - called *neurons*. The neurons are linked through *connections*; the connections are directed, and with each connection, there is an associated *weight*.

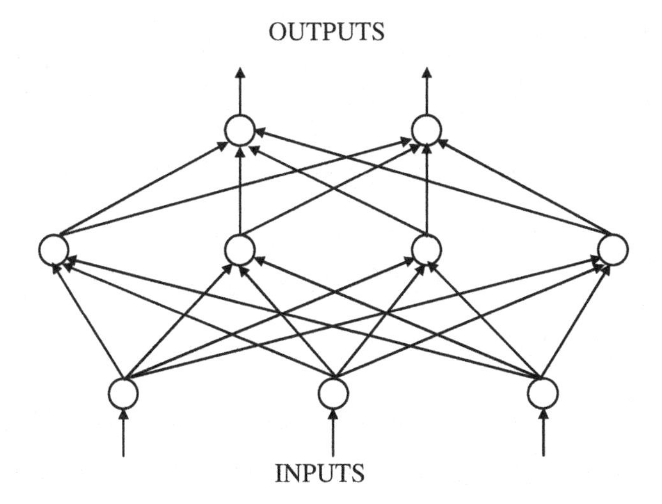

Fig. 1.2 Exemplary three-layer neural network

The first simplified model of the neuron was introduced by McCulloch and Pitts in 1943. The model is described in Fig. 1.3.

The neuron is a simple processor with the following elements:

the inputs

the external signals coming to the neuron; denoted by x_i, where $i = 1, 2, ..., N$ is the index of the incoming signal, N is the number of incoming signals; while the inputs associated with the external inputs to the network and the inputs related to the connections between neurons within the network are distinguished,

the weights

the values describing the strength of each connection; denoted by w_{ij}, where $i = 1, 2, ..., N$ is the index of the incoming signal, and j is the index of the considered neuron,

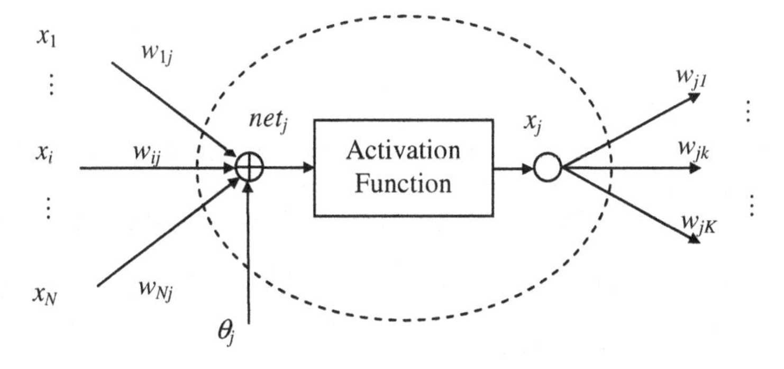

Fig. 1.3 Elementary model of a neuron

the summing points

the simple functions collecting all the inputs incoming to each neuron, taking into account weighting of the connections, described by the function net_j

$$net_j = \sum_{i=1}^{N} w_{ij}\, x_i + \theta_j \tag{1.1}$$

where θ_j is a *bias* weight, if it is written $net_j = \sum_{i=1}^{N} w_{ij}\, x_i$ we tacitly understand that the bias is included,

the activation functions

the basic functions transforming the functions net_j to the output of the neuron; there are several different activation functions $f(net_j)$ used in the neural networks, in this book we will consider only two kinds of them, which are characterized by the existence of smooth derivatives and are easily to linearized e.g. (Rumelhart et al. 1986, Anderson and Rosenfeld 1989, Werbos 1989, Hornik 1989, Dreyfus 1990, Tadeusiewicz 1993, Korbicz et al. 1994, Duch et al. 1999). In the next two figures, these two activation functions are shown. In Fig. 1.4, a *unipolar* sigmoidal activation function is pictured, which has the following analytical form (Hertz 1990, Anderson and Rosenfeld 1989, Amari 1993, Hassoun 1995, Fine 1999, Krawczak 2003e)

$$f(net, \lambda) = \frac{1}{1 + \exp(-\lambda\, net)}. \tag{1.2}$$

This function has particularly significant properties, namely a change of the parameter λ (sometimes $1/\lambda$ is called the temperature of the neural system (Hertz et al. 1990)) causes a change of the function in this way that for $\lambda \to \infty$ the unipolar activation functions becomes the step function (or the Heaviside function), which was considered by McCulloch and Pitts (1943)

$$f(net, \infty) = \begin{cases} 1 & \text{if } net \geq 0 \\ 0 & \text{otherwise} \end{cases} \tag{1.3}$$

For $\lambda \to 0$ the activation function (1.2) becomes more and more linear, here for λ being a small positive value we will call this function "almost" linear; there is another activation function of similar properties, it is the *bipolar* one, e.g. (Zurada 1992, Ellacott 1993), see Fig. 1.5, of the following analytical form

$$f(net, \lambda) = \frac{2}{1 + \exp(-\lambda\, net)} - 1. \tag{1.4}$$

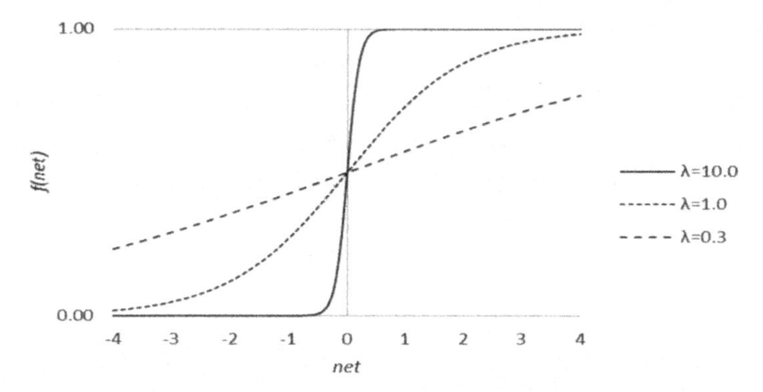

Fig. 1.4 Unipolar sigmoidal neuron's activation function

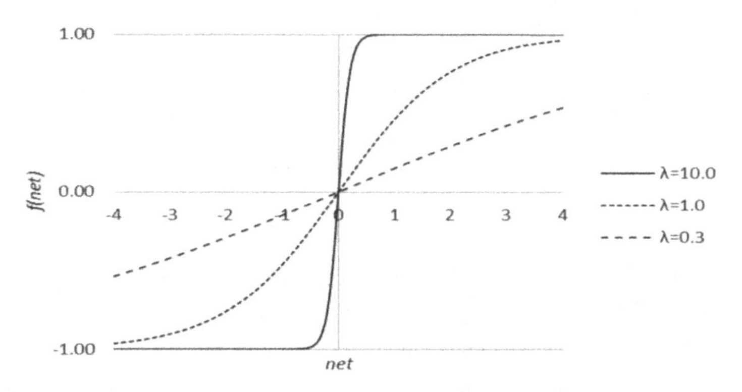

Fig. 1.5 Bipolar sigmoidal function

The changing of parameter λ is a base of simulated annealing technique used for solving nonlinear problems e.g. (Kirkpatrick et al. 1983, Aarts and Laarhoven 1987, Brooks and Morgan 1995) as well as a base to the homotopy method (Richter and de Carlo 1984, Krawczak 2000c, Krawczak 2000e) or the continuation method (Avila 1974, Krawczak 2000b, Krawczak 2000d).

The other aspects of the neuron are:

the state of the neuron, or *the output of the neuron*
the present value of the activation function of the neuron; $k = 1, 2, ..., K$ in Fig. 1.3 denotes the number of outcoming connections,

the junction points
the state of each neuron is transmitted to the neurons connected to the considered one with exactly the same value.

The neurons constituting the neural network are arranged in layers. We will label the layers with the index l, where $l = 1, 2, ..., L$. A layer consists of neurons whose number will be denoted by $N(l)$. In order to describe the neuron's connections in a unique way, the following notation is introduced (a little different than in common literature). Let us consider the neuron indexed by $j(l)$, associated with the l-th layer, $l = 1, 2, ..., L$. Such neuron's position in the network will be described by $j(l)$. Now, this considered neuron located in the l-th layer of the network is depicted as follows (Krawczak 2003b, Krawczak 2003e), see in Fig. 1.6.

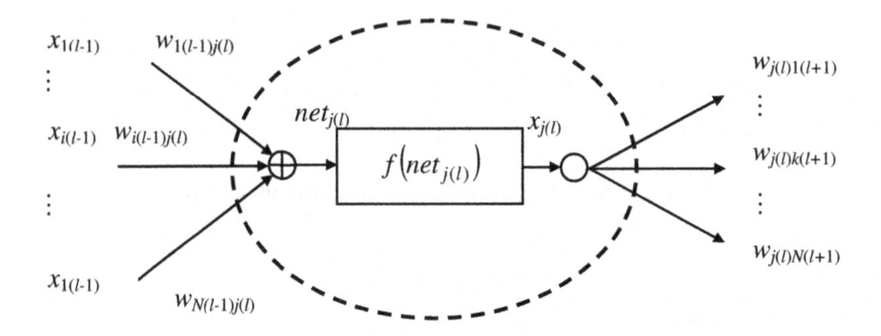

Fig. 1.6 The neuron with described position in the network

The neuron inputs are denoted $x_{i(l-1)}$, here $i(l-1) = 1(l-1), 2(l-1), ..., N(l-1)$ for $l = 1, 2, ..., L$, for simplicity we will write $i(l-1) = 1, 2, ..., N(l-1)$; the weights connecting the $j(l)$-th neuron associated with the l-th layer, $l = 1, 2, ..., L$, with other neurons located in the $(l-1)$-st layer are denoted by $w_{i(l-1)j(l)}$ - the subscript $i(l-1)j(l)$ describes the weight of the output of the i-th neuron belonging to the $(l-1)$-th layer with the neuron j of the l-th layer. If there are weights $w_{i(l)j(l)}$ then the network is called *recurrent*, and in the considered case, the feedforward becomes the *multilayer* one. The result of summing within the $j(l)$ neuron is as follows:

$$net_{j(l)} = net_j(l) = \sum_{i(l-1)=1}^{N(l-1)} w_{i(l-1)j(l)}\, x_{i(l-1)} + \theta_{j(l)} \tag{1.5}$$

the activation functions (1.2) or (1.4) with the usually assumed $\lambda = 1.0$ now has the forms

$$f(net_{j(l)}) = \frac{1}{1 + \exp(-net_{j(l)})} \tag{1.6}$$

$$f(net_{j(l)}) = \frac{2}{1 + \exp(-net_{j(l)})} - 1 \tag{1.7}$$

equal to the state of the neuron or the output of the neuron.

The signals external to the network and coming into it are called the *input* to the neural network. They feed the network via the first layer, $l = 1$. All the inputs are collected within the extra layer indexed by $l = 0$ of dimension $N(0)$.

The output of the network is equivalent to all the neurons' outputs from the last L-th layer. The network output is strictly related to the presented input, subject to the conditions resulting from the constancy of the structure (the neuron connections), the activation functions as well as the weights. In this way the neural networks realize the following *simulation*, that is

$$output_p = NN(input_p) \tag{1.8}$$

where the index p is introduced in order to distinguished different inputs and related outputs. Due to the nonlinearity of the activation functions (1.1) or (1.4), the mapping (1.8) is also nonlinear as well as not unique.

1.3 Neural Networks Learning Process

In this section, we will deal with the basic properties of learning or training process of the neural networks. First, we select the *architecture* of the network, meaning the number of layers L and the numbers of neurons $N(l)$, $l = 1, 2, ..., L$ located in each layer. The architecture is selected especially for the considered problem, which determines the number of inputs $N(0)$ and the number of neurons in the last layer $N(L)$. Including the inputs we can consider the following indices of layers $l = 1, 2, ..., L$.

Here, we consider the supervised learning paradigm, which can be stated as follows. There is a set of examples (or patterns) available; each example is a pair of inputs and desired outputs $\{input_p, output_p\}$, $p = 1, 2, ..., P$, where P denotes the number of examples available. The aim of learning is to adjust the weights of connections between the neurons

$$w_{i(l-1)j(l)},\ i(l-1) = 1, 2, ..., N(l-1),\ j(l) = 1,\ 2, ..., N(l),\ l = 1, 2, ..., L,$$

according to some predefined rules. The fundamental rule for learning is the *Hebbian learning rule*, invented by Hebb (1949), that states:

If two neurons are active in the same time then their strength of connection increases.

The simplest mathematical formulation of the Hebb's learning rule is related to the modification of weight values according to the formula, e.g. (Zurada 1992, 1996),

$$\Delta w_{ij} = \eta\ x_i\ x_j \tag{1.9}$$

where η is a positive number called *learning rate*, while x_i and x_j are states of two neurons.

The most used formulation of the Hebb's learning rule, belonging to Widrow and Hoff (1960), is called the *delta rule* and regards the difference between the actual and desired states for the modification of the weights

$$\Delta w_{ij} = \eta \left(d_j - x_j\right) x_i . \tag{1.10}$$

This idea is the foundation of almost all learning algorithms for neural networks, see subsequent chapters of this book.

The performance index of learning (often called the cost of learning or the error of learning) gives a functional form of the quality of learning, that is, how exactly the output of the network fits the desired output. The performance index describes a measure of difference between the output of the network and the desired output.

All the learning algorithms adjust the weights of the trained network in order to minimize the chosen performance index. In the literature, e.g. Duch et al. (1999), we can find different measures of fitting the output of the network to the desired output. We can distinguish two classes of such measures. The first is the *quadratic* performance index

$$E = \frac{1}{2}\sum_{p=1}^{P}\left(D_p - X_p(L)\right)^2 = \frac{1}{2}\sum_{p=1}^{P}\sum_{i(L)=1}^{N(L)}\left(d_{pi(L)} - x_{pi(L)}\right)^2 . \tag{1.11}$$

where $D_p = \left[d_{p1}, d_{p2},..., d_{pN(L)}\right]^T$ is the desired vector of outputs, and the response of the network is $X_p(L) = \left[x_{p1}, x_{p2},..., x_{pN(L)}\right]^T$, P is the number of the learning examples, while the second is the following *entropy* performance index

$$E = \frac{1}{2}\frac{1}{P}\sum_{p=1}^{P}\sum_{i(L)=1}^{N(L)}\left(\left(1 + d_{pi(L)}\right)\log\frac{1 + d_{pi(L)}}{1 + x_{pi(L)}} + \left(1 - d_{pi(L)}\right)\log\frac{1 - d_{pi(L)}}{1 - x_{pi(L)}}\right). \tag{1.12}$$

Both performance indices are continuous and differentiable.

1.4 Generalization

The training set consists of pairs of examples $\left\{D_p, X_p(L)\right\}$, $p = 1,2,..., P$. After learning, the neural network can exactly simulate the desired output D_p as the response of the network to the input $X_p(0)$. We would like the network not only to learn the mapping for the pairs from the training set, but also to generate reasonable outputs for inputs from outside of the training set. Such a problem is called the *generalization problem* (e.g. Hecht-Nielsen 1990, 1992). The generalization is essential both for classification and approximation problems. In the first case, the evaluation of generalization can be done only by dividing the training set into two parts, so as to train the network using one part and to test the generalization using the second part of the training set. In the approximation case, the generalization is evaluated by comparing both the functions and their derivatives.

1.5 Foundation of Neural Networks

From the beginning of the neural networks history, the fundamental question was if and how the neural networks could realize functional forms. The first idea revolves around implementing Boolean functions, like e.g. EXCLUSIVE OR (Rashevsky 1948, Minsky and Papert 1969) by neural networks.

There are some works, e.g. Hecht-Nielsen (1990), suggesting that the Kolmogorov's theorem (1957), concerning the realization of any multivariate function by superposition of other continuous functions with single variables, could be applied to the neural networks. Meanwhile, though, there is no formal proof that the Kolmogorov's theorem can be directly applied for proving the universality of the neural networks as universal function approximators, not even in the paper by Kurkova (1992) supporting the relevance of this theorem to the neural networks realization.

It is recognized that the most significant results supporting the proposition that the multilayer neural networks are the *universal approximators* were given by Cybenko (1989), and independently by Hornik, Stinchcombe and White (1989), and Funahashi (1989).

Cybenko's Theorem, based on the Hahn-Banach theorem, and following the descriptions of Fig. 1.7, can be summarised as follows (Cybenko 1989).

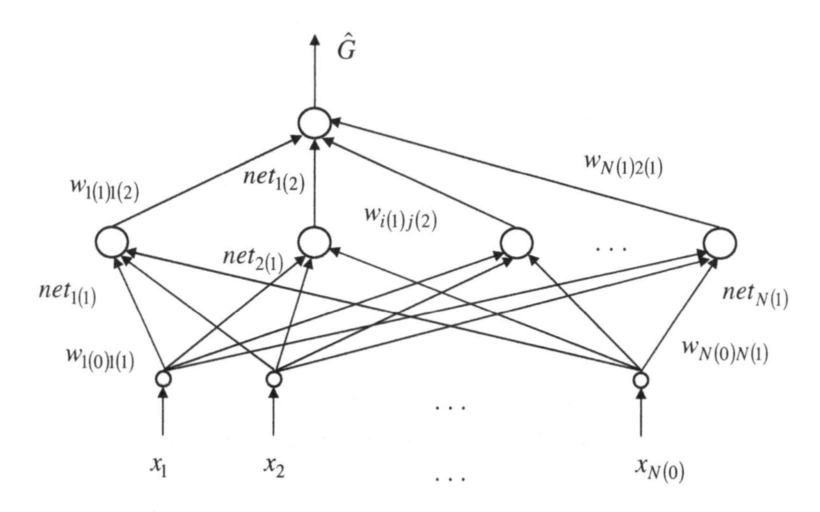

Fig. 1.7 Approximation of G by the two-layer neural

There is given any continuous real-valued function $G(x)$ on $[0, 1]^{N(0)}$ (or any compact subset of $R^{N(0)}$), where $x = [x_1, x_2, ..., x_{N(0)}]^T$, and $\varepsilon > 0$. For any continuous sigmoidal-type function, e.g.

$$f(net_{i(1)}) = \frac{1}{1 + e^{-net_{i(1)}}}, \quad i(1) = 1, 2, ..., N(1) \tag{1.13}$$

and $f(net_{1(1)}) = net_{1(1)}$ there exist vectors

$$w = [w_{1(0)1(1)}, w_{2(0)1(1)}, ..., w_{N(0)1(1)}, w_{1(0)2(1)}, w_{2(0)2(1)}, ..., w_{N(0)2(1)}, ..., w_{1(0)N(1)}, ..., w_{N(0)N(1)}]^T$$

and

$v = [w_{1(1)N(2)}, w_{2(1)N(2)}, ..., w_{N(1)N(2)}]^T$, and $\theta = [\theta_{j(1)}, \theta_{j(2)}, ..., \theta_{j(N(1))}]^T$, and a parameterised function $\hat{G}(x, w, v, \theta)$: $[0, 1]^{N(0)} \rightarrow R$ such that

$$\left|\hat{G}(x, w, v, \theta) - G(x)\right| < \varepsilon \quad \text{for all } x \in [0, 1]^{N(0)} \tag{1.14}$$

where

$$\hat{G}(x, w, v, \theta) = \sum_{j(1)=1}^{N(1)} w_{j(1)1(2)} \, f\left(\sum_{i(0)=1}^{N(0)} w_{i(0)j(1)} \, x_{i(0)} + \theta_{j(1)}\right) \tag{1.15}$$

and

$$w_{i(0)j(1)} \in R, \; i(0) = 1, 2, ... N(0), \; j(1) = 1, 2, ... N(1),$$
$$w_{j(1)1(1)} \in R, \; \theta_{j(1)} \in R, \; j(1) = 1, 2, ... N(1);$$

in (1.13)

$$net_{j(1)} = \sum_{i(0)=1}^{N(0)} w_{i(0)j(1)} \, x_{i(0)} \text{ and } net_{1(1)} = \sum_{j(1)=1}^{N(1)} w_{j(1)1(2)} \, x_{j(1)},$$

where $x_{j(1)}$ denotes the $j(1)$-th neuron output, $j(1) = 1, 2, ... N(1)$.

Equ. 1.15 gives the neural network implementation shown in Fig. 1.7, where three layers can be distinguished:

- the 0-th layer is responsible for summing the inputs subject to the weights $w_{i(0)j(1)} \in R$, $i(0) = 1, 2, ... N(0)$, $j(1) = 1, 2, ... N(1)$,
- the 1-st layer of dimension $N(1)$ is responsible for nonlinear components of the approximation,
- the 2-nd layer (the output layer) is responsible for summarizing the components with respect to the weights $w_{j(1)1(1)}$, $j(1) = 1, 2, ... N(1)$.

In Equ. 1.15, we can notice that double summation structure is required to approximate the function G with prescribed accuracy ε. In Fig. 1.7, the network can be treated as a two-layer with one extra layer collecting the inputs $x_{i(0)}$, $i(0)=1,2,...,N(0)$. In the neural networks literature, such a network is said to be three-layer with one hidden layer (not visible from the outside of the network).

The conclusion of the Cybenko's theorem is the following, namely any continuous function G can be approximated, with predefined accuracy, by a three-layer neural network with sigmoidal activation functions.

The results of Hornik, Stinchcombe and White (1989), based on the Stone-Weierstrass theorem, and the results of Funahashi (1989), based on the integral formula by Irie and Miyake (1988), established that three-layer neural network (with one hidden layer) could approximate any continuous multivariate function with a prescribed accuracy. Other results obtained among others by Ito (1991), Sun and Cheney (1992), Light (1992) and Hornik (1993), Katsuura and Sprecher (1994) extended Cybenko's theorem to different shapes of the activation functions.

Cybenko's theorem can be extended (Cybenko 1989, Hasaun 1995) for the *universal classifiers* and takes the following form. For any real-valued function

$$G(x)=j \quad \text{if and only if } x\in P_j \tag{1.16}$$

where $x=\left[x_1,x_2,...,x_{N(0)}\right]^T$, and $G(x): A^{N(0)} \to \{1, 2,..., k\}$, $A^{N(0)}$ is a compact (closed and bounded) subset of $R^{N(0)}$, and $P_1, P_2,..., P_k$ is a partition of $A^{N(0)}$ into k disjoint measurable subsets, i.e. $A^{N(0)}=\bigcup_{j=1}^{k} P_j$ and $P_{j1}\cap P_{j2}$ is empty for $j1\neq j2$, and $\varepsilon>0$, there exist vectors

$w=\left[w_{1(0)1(1)}, w_{2(0)1(1)},..., w_{N(0)1(1)}, w_{1(0)2(1)}, w_{2(0)2(1)},..., w_{N(0)2(1)},..., w_{1(0)N(1)},..., w_{N(0)N(1)}\right]^T$ and $v=\left[w_{1(1)N(2)}, w_{2(1)N(2)},..., w_{N(1)N(2)}\right]^T$, and $\theta=\left[\theta_{j(1)},\theta_{j(2)},...,\theta_{j(N(1))}\right]^T$, and a parameterised function $\hat{G}(x, w, v, \theta)$: $A^{N(0)}\to\{1, 2,..., k\}$ such that

$$\left|\hat{G}(x, w, v, \theta)-G(x)\right|<\varepsilon \quad \text{for all } x\in P_j \tag{1.17}$$

where

$$\hat{G}(x, w, v, \theta)=\sum_{j(1)=1}^{N(1)} w_{j(1)1(2)}\, f\left(\sum_{i(0)=1}^{N(0)} w_{i(0)j(1)}\, x_{i(0)}+\theta_{j(1)}\right) \tag{1.18}$$

and

$$w_{i(0)j(1)}\in R,\ i(0)=1,2,...N(0),\ j(1)=1,2,...N(1),$$
$$w_{j(1)1(1)}\in R,\ \theta_{j(1)}\in R,\ j(1)=1,2,...N(1);$$

in (1.18)

$$f\left(net_{i(1)}\right)=\frac{1}{1+e^{-net_{i(1)}}},\ i(1)=1,2,\ldots,N(1) \tag{1.19}$$

$$net_{j(1)}=\sum_{i(0)=1}^{N(0)} w_{i(0)j(1)}\, x_{i(0)} \text{ and } net_{1(1)}=\sum_{j(1)=1}^{N(1)} w_{j(1)1(2)}\, x_{j(1)},$$

where $x_{j(1)}$ denotes the $j(1)$-th neuron output, $j(1)=1,2,\ldots N(1)$.

Again, the conclusion is the same as before, namely that three-layer neural network (with one hidden layer) with sigmoidal activation functions is a universal classifier.

In order to treat a neural network as a universal approximator or classifier, we must be able to find the weights

$$w_{i(0)j(1)} \in R,\ i(0)=1,2,\ldots N(0),\ j(1)=1,2,\ldots N(1),$$

$$w_{j(1)1(1)} \in R,\ \theta_{j(1)} \in R,\ j(1)=1,2,\ldots N(1)$$

as well as the number of the hidden neurons $N(1)$.

1.6 Milestones of Artificial Neural Networks

In this section, we will list the publications, which are the most important, according to the author of this book, in the neural networks field.

The history of the neural network systems began with the paper by W. McCulloch and W. Pitts in 1943 (McCulloch and Pitts 1943). They introduced the first mathematical model of a neuron as a basic processing element in a network. The model was a unimodal step function with a threshold. They showed that a network with extremely basic elements (neurons) could compute an arbitrary logical or arithmetical function.

In 1938 N. Rashevsky (Rashevsky 1938) forwarded the hypothesis that the brain's functionality could be based on binary logic operations.

In 1947 W. Pitts and W. McCulloch (Pitts and McCulloch 1947) in a rather forgotten and philosophical text considered a hypothetical algorithm for learning, using e.g. names like an *apparition* as patterns, or *computor* as a machine for computing. The present-day supervised learning paradigms are remarkably close to their ideas.

In 1949 D. Hebb in a book *The Organization of Behaviour* proposed a hypothetical way of stimulating the neurons: "*Let us assume then that the persistence of repetition of a reverberatory activity (or trace) tends to induce lasting cellular changes that add to its stability. The assumption can be precisely stated as follows: When an axon of cell A is near enough to excite cell B and repeatedly or persistently takes part in firing it, some growth process or metabolic change takes place on one or both cells so that A's efficiency as one of the cells firing B is increased*", (Hebb 1949). This learning paradigm is known as the *Hebbian rule*.

F. Rosenblatt (1958) studied the McCulloch neuron models and the Hebb prescription of neurons' excitement. The term *perceptron* became common in use due to the book *Principles of Neurodynamics,* published in 1962. Rosenblatt (1962) proved that training of a two-layer network, used for linear separability of two sets, can be performed iteratively and converges in a limited number of steps.

In 1960 B. Widrow and M. Hoff published a paper in which they introduced the ADAptive LINear Element, called ADALINE, which was the McCulloch and Pitts neuron model with an interesting way of weight adjustment. Widrow and Hoff (1960) proposed a new algorithm for learning, which used a difference between the computed and desired outputs. The methodology, similar to the idea of Pitts and McCulloch from 1947, was named the *delta rule* and is fundamental in substantial learning algorithms.

The first period of the neural network field was closed by a book *Perceptron* by M. Minsky and S. Papert (1969). In this book they showed that a single layer network is not capable of solving the nonlinear separable problems. They showed that only the extension of the number of layers could solve this class of problems, although they also said that the problem of training of multilayer networks could be unsolvable. This statement caused that the interest in the field was practically stopped.

At the beginning of the 1980s, the study of the neural networks started again due to a paper by J. Hopfield (1982) and by J. Hopfield and D. Tank (1985). Hopfield demonstrated connections between some models of the neural networks and physical models of symmetric connections and a spin glass. Hopfield introduced a new single layer recurrent architecture of neural networks, now known as the *Hopfield nets*. The work was continued.

In 1982 T. Kohonen described self-organization of neural networks. This class of neural networks uses the fact that sensory signals are represented as two-dimensional *images* in the brain. Kohonen (1982, 1987, 1997) introduced a mechanism using a neighbourhood function, which allows clustering of similar input signals.

G. Hinton and T. Sejnowski published in 1983 a paper, in which they introduce a neural network called the *Boltzmann machine*. They extended the Hopfield network by adding stochastic dynamics. Hinton and Sejnowski (1983) also used sigmoidal neuron models. (It seems to be significant due to the later Hinton's collaboration with Rumelhart).

In 1986 D. Rumelhart, G. Hinton and R. Williams published a paper in which they described a new learning algorithm for multilayer neural networks. The algorithm is called *the backpropagation* due to the reverse direction of the learning error signal in comparison with the signal propagated from the input to the output. They (Rumelhart et al. 1986) used sigmoidal neuron models and applied a version of the gradient algorithm for weight adjustment. The backpropagation algorithm opened new possibilities of investigating and applying the multilayer neural networks.

The list is certainly not complete at all; there are many other eminent scientists whose significant publications contributed to the creation and development of the field of neural computation.

Chapter 2
Basics of Generalized Nets

2.1 Introduction

The concept *generalized nets* was described by Krassimir Atanassov in 1982.

Since that time hundreds papers and several books have been published. In *Review and bibliography on generalized nets theory and applications* by Radeva, Krawczak and Choy (2002), one can find the list of 353 scientific works related to the generalized nets. In the book *A Survey of Generalized Nets,* there is one chapter entitled *Review and bibliography on generalized nets theory and applications* by Alexieva, Choy and Koycheva (2007) with a bibliography of 638 works about generalized nets. Nowadays, there is the http://www.ifigenia.org/ web resource which collects events and publication in the areas of intuitionistic fuzzy sets (IFS) and generalized nets (GN).

First of all, it must be emphasised that Atanassov, in many papers (e.g. 1984, 1987, 2005, 2006), and several books (1991, 1992, 1997, 1998, 2007), as a co-author (Atanassov and Aladjov 2000, Atanassov et al. 2006, Krawczak et al. 2010), or as an editor (Kacprzyk et al. 2005, Atanassov et al. 2006, 2007a, 2007b, 2008a, 2009, 2010, 2011, 2012), proposed a new definition of nets for describing and analysing various kinds of discrete events systems.

A large part of publications related to the generalized nets can be found at http://www.daimi.aau.dk/PetriNets/bibl/aboutpnbibl.html

Generalized nets are defined in a different way than nets introduced by Carl-Adam Petri (1962, 1980), Genrich and Lautenbach (1979), Starke (1980), Jensen (1981), and Etzion and Yoeli (1983). The idea of generalized nets is based on developing the relation of *place* and *transition.*

Generalized nets have some new elements, and the elements of other nets can be described in terms of generalized nets. For example, the tokens in generalized nets do not have colours (as in the coloured Petri nets, e.g. Zerros and Irani 1997, Jensen 1981), generators of random numbers (as in the stochastic Petri nets, e.g. Shapiro 1979), or inhibitor arcs (as in the super nets). It has been proved (Atanassov 1991, 2007) that generalized nets allow comparing various kinds of nets.

Similarly to other kinds of nets, the generalized nets have *static structure, dynamic elements* (*tokens*) and *temporal elements.* Generalized nets have three global temporal constants:

M. Krawczak: *Multilayer Neural Networks*, SCI 478, pp. 15–30.
DOI: 10.1007/978-3-319-00248-4_2

- o the initial moment in which the net starts functioning,
- o the elementary time-step of the process,
- o the duration of functioning

while other kinds of nets can start functioning at any moment.

The static structure of generalized nets is characterized by *transitions* and is the same as in the E-nets or the generalized E-nets.

The dynamic elements - *tokens* - in generalized nets are described by *characteristics,* which is changeable. The characteristics of tokens are responsible for carrying information and form a sort of *memory* of the nets.

We can point out the following main reasons to introduce the generalized nets:

- o comparing different types of nets as mathematical objects,
- o investigating general properties of the generalized nets and transfer them to other nets,
- o analysing details of processes.

Similarly to the theory of Petri nets, the theory of generalized nets can be divided into two basic areas, namely *the special theory of generalized nets* and *the general theory of generalized nets.*

The special theory of generalized nets is related to the definitions and the properties as well as modifications of generalized nets. The special theory of generalized nets consists of problems of conflicts avoidance or the transfer of definitions and properties of Petri nets e.g. the properties: *purity*, *boundedness*, *liveness*, *conservation* and others, or properties of reduced generalized nets (there are some removed elements). There are known some extensions of generalized nets, like fuzziness of the truth-values of predicates, the colour generalized nets, the generalized nets with a complex graphical structure.

The general theory of generalized nets can be assessed in different aspects: algebraic, logical, operation, program, methodological and topological.

There are additional elements in generalized nets but they provide more and greater modelling possibilities and determine the place of generalized nets among the different kinds of Petri nets.

The succeeding sections of this chapter are based on books by Atanassov (1991, 1992, 1997, 1998, 2007) as well as a book by Krawczak (2003e) and by Krawczak et al. (2010).

2.2 Concept of Generalized Nets

The first basic difference between generalized nets and the standard Petri nets is the relation between the objects called *places* and the objects called *transitions* (Atanassov 1991), this relation has a rather complex nature. Any place is marked by a circle O while any transition by a bar with a triangle ▼(strictly speaking, a graphic structure of the place and transition). Generalized nets contain *tokens*, each token can be treated as a granule of information transferred from one place to another. The information bearing by the token is described by *token's characteristic*. At the beginning, each token before functioning within the generalized net has some starting information referred to as *an initial characteristic* here. The token can be transferred from one place to another, passing the transition. During staying or rather passing the transition the token's characteristics is modified – it can be understood that carrying information by the token is changed within some conditions involved in the transition while the token passes the transition – the granule of information is modified.

There are distinguishable two kinds of places, namely *input places* and *output places* (see Fig. 2.1).

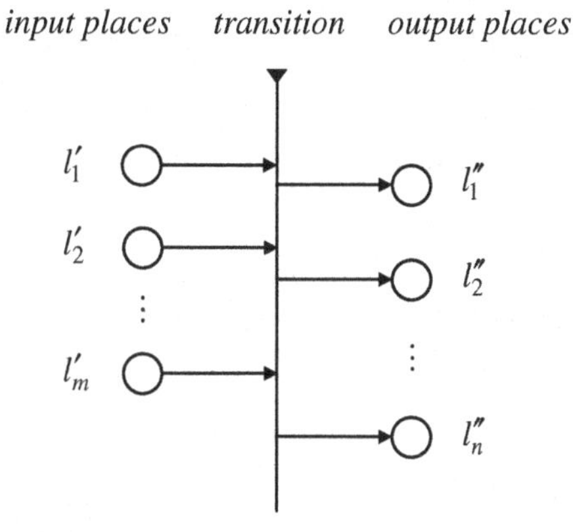

Fig. 2.1 Transition with input and output places

Formally, every transition is described by a seven-tuple

$$Z = \langle L', L'', t_1, t_2, r, M, \square \rangle \tag{2.1}$$

where:

$L' = \{l'_1, l'_2, \ldots, l'_m\}$
is a finite and non-empty set of the input places of the transition,

$L'' = \{l''_1, l''_2, \ldots, l''_m\}$
is a finite and non-empty set of the output places of the transition,

t_1
is the present time of firing of the transition,

t_2
is the duration of the transition being in active,

r
is *the condition* which determines which tokens are allowed to pass or to be transferred via the transition from the prescribed input place to the prescribed output place. The condition r has the form of *an index matrix* described in Atanassov (1987) and can be shown in the following way:

$$r = \begin{array}{c|ccccc} & l''_1 & \cdots & l''_j & \cdots & l''_n \\ \hline l'_1 & r_{11} & \cdots & r_{1j} & \cdots & r_{1n} \\ \vdots & \vdots & \cdots & \vdots & \cdots & \vdots \\ l'_i & r_{i1} & \cdots & r_{ij} & \cdots & r_{in} \\ \vdots & \vdots & \cdots & \vdots & \cdots & \vdots \\ l'_m & r_{m1} & \cdots & r_{mj} & \cdots & r_{mn} \end{array}$$

where r_{ij}, $i = 1,2,...,m,\quad j = 1,2,...,n,$ is a predicate that corresponds to the i-th input place and the j-th output place. When $r_{ij} = true$, for some set values i and j the token is allowed to pass the transition from the i-th input place to the j-th output place,

M
is an index matrix of the capacities of arcs of the transition, it means a number of tokens simultaneously transferred via the transition:

$$M = \begin{array}{c|ccccc} & l''_1 & \cdots & l''_j & \cdots & l''_n \\ \hline l'_1 & m_{11} & \cdots & m_{1j} & \cdots & m_{1n} \\ \vdots & \vdots & \cdots & \vdots & \cdots & \vdots \\ l'_i & m_{i1} & \cdots & m_{ij} & \cdots & m_{in} \\ \vdots & \vdots & \cdots & \vdots & \cdots & \vdots \\ l'_m & m_{m1} & \cdots & m_{mj} & \cdots & m_{mn} \end{array}$$

where $m_{ij} \geq 0$, $i = 1,2,...,m,\quad j = 1,2,...,n,$ are natural numbers,

$\square$
is an expression of a Boolean form, it may contain as variables the symbols that describe labels of the input places, the expression $\square$ is constructed from variables and the Boolean connectives $\wedge$ and $\vee$, the semantics of which is defined in the following way:

$\wedge(l_{i1}, l_{i2}, \ldots, l_{iu})$
every place $l_{i1}, l_{i2}, \ldots, l_{iu}$ must contain at least one token,

$\vee(l_{i1}, l_{i2}, \ldots, l_{iu})$
there must be at least one token in every place $l_{i1}, l_{i2}, \ldots, l_{iu}$,
where $\{l_{i1}, l_{i2}, \ldots, l_{iu}\} \subset L'$,

when the value of a Boolean a type expression □ is *true*, it means that the transition can become active, otherwise it cannot.

The formal definition of *the generalized net* can be stated as an ordered four-tuple:

$$E = \left\langle \langle A, \pi_A, \pi_L, c, f, \Theta_1, \Theta_2 \rangle, \langle K, \pi_k, \Theta_K \rangle, \left\langle T, t^0, t^* \right\rangle, \langle X, \Phi, b \rangle \right\rangle \tag{2.2}$$

where each tuple contains several elements described as follows:

A
is a set of transitions,

π_A
is a function giving the priorities of the transitions, i.e. $\pi_A : A \to N$, where $N = \{0, 1, 2, \ldots\} \cup \{\infty\}$,

π_L
is a function giving the priorities of the places, i.e. $\pi_L : L \to N$, where

$$L = pr_1 A \cup pr_2 A$$

for the set of all the generalized net places L,

c
is a function describing the capacities of the places, i.e. $c : L \to N$,

f
is a function that calculates the values (*true* or *false*) of the predicates of the transition's conditions (instead of values *true* or *false* we can use values from the set $\{0, 1\}$),

Θ_1
is a function specifying the next moment for a given transition Z that can be activated, i.e.

$$\Theta_1(t) = t', \text{ where } pr_3 Z = t, \ t' \in [T, T + t^*] \text{ and } t \le t',$$

this value is calculated just at the moment when functioning of the transition stops,

Θ_2
is a function yielding the period for activation state of the transition Z, i.e.

$$\Theta_2(t) = t', \text{ where } pr_4 Z = t \in [T, T + t^*] \text{ and } t' \ge 0,$$

this value of the function is calculated at the beginning of the transition's functioning,

K
is the set of the tokens of the generalized net; in some cases,

π_K
is a function describing the priorities of the tokens, i.e. $\pi_K : K \to N$,

Θ_K
is a function which generates the moment when a given token can enter the net, i.e.

$$\Theta_K(\alpha) = t \text{, where } \alpha \in K \text{ and } t \in [T, T + t^*],$$

T
is the starting moment of functioning of the generalized net, the moment has a global character in time scale,

t^0
is an elementary sampling period related to the global time scale,

t^*
is the period of functioning of the generalized net,

X
is the set of all initial characteristics of the tokens entering the net (related to initial information possessed by each token),

Φ
is a characteristic function which determines the new characteristics of each token; the function is obtained during passing the given transition from the input place to the output place (it is responsible for changing information carried by tokens when they pass the proper transition),

b
is a function that determines the number of changing characteristics by tokens (separately for each token), i.e. $b: K \to N$, and for example, $b(\alpha) = 1$ means that the token α enters the net with some initial characteristics and they will not be changed the generalized net functioning, in the case of $b(\alpha) = k < \infty$ the token α will keep its last k characteristics (in other words, the token memorises the last k changes of carried information, while for $b(\alpha) = \infty$ the token α will keep all its characteristics obtained during the generalized net functioning time.

The number of elements of the description of generalized nets (2.2) seems to be rich and, at the beginning, reading is troublesome and one can face some difficulties when getting accustomed to this rather new nomenclature. Fortunately, most of the considered discrete time systems do not require the use of the whole elements, in such cases any missing element can be omitted. The generalized nets with some disused elements are called the *reduced generalized nets*.

As was mentioned before, within the description of a generalized net we can distinguish a part related to the static and the dynamic structure. The static structure of a given generalized net is determined by the following elements:

- o input places and the output places,
- o index matrices of the arcs and the transition type,
- o functions π_A, π_L related to priorities of transitions and places as well c related to capabilities of places

for each transition.

The dynamic structure of the generalized nets is related to:

- o tokens (carriers of information),
- o conditions of transitions,
- o functions f calculating values of the predicates of conditions and π_K describing the priorities of tokens,
- o temporal elements related to time T, t^0, t^* and Θ_1, Θ_2 and Θ_K.

2.3 Algorithm of Generalized Nets

It seems that the definition of a generalized net is more general and therefore more complicated in comparison to the standard Petri net. Therefore, the transferring algorithms of the tokens within a generalized net must be also more complex. Generalized nets are more general, therefore, the algorithms responsible for movement of tokens cannot be as simple as in Petri nets. For example, in a Petri net implementation the functionality of parallelism is reduced to a sequential activation of transitions or depends on the priorities of transitions. In the case of generalized nets movements of tokens, described by proper algorithms, the movement depends on priorities of places, priorities of transitions as well as priorities of tokens. These give the possibilities to describe functioning of a dynamic system much more detailed.

Here we will describe the general algorithm of transferring tokens through a transition in generalized nets. It is assumed that the general algorithm starts at the moment $t_1 = TIME$ and can be described in the following way:

STEP 1:
Divide the tokens into two groups:

1) group P_1 of tokens, which can be directly transfer from the input places via the transition to the output places,
2) group P_2 of tokens, which will be processed passing the transition.

STEP 2:
Sort the input places of the transitions by their priorities.

STEP 3:
Sort the output places of the transitions by their priorities.

STEP 4:
Sort the tokens from group P_1 of the input places by their priorities (providing results of STEP 2), create the index matrix R on the base of the index matrix r in the following way:

$$R_{ij} = \begin{cases} 1 & \text{if } r_{ij} = true \\ 0 & \text{if } r_{ij} = false \text{ or if the value is determined in STEP 5} \end{cases}$$

for $i = 1, 2, , m$, $j = 1, 2, ... n$.

STEP 5:
Assign value 0 to all elements of R for the following cases:

1) the group P_1 is empty,
2) the output place which corresponds to the respective predicate exceeded the output place capacity,
3) the capacity of the arc between the corresponding input and output places is 0.

STEP 6:
Assign the values of R before calculating other elements of the index matrix r.

STEP 7:
Calculate the values of the characteristic functions related to the tokens which will enter the output places.

STEP 8:
Considering the input place priorities, for each input place follow the procedure:

1) sort the tokens with respect to the highest priority in the input place,
2) for group P_2 transfer the tokens to all output places for which the corresponding predicate permits.

STEP 9:
Update the group P_2 by transferring the tokens with values of predicates equal to *false* in sequence of the highest priority or the tokens which cannot be transferred to the corresponding output places due to exceeding the capacities of the output places.

STEP 10:
Change the time: $TIME := TIME + t^0$.

STEP 11:
If the time $TIME := TIME + t^0 < t_1 + t_2$ go to STEP 4.

STEP 12:
If the time $TIME := TIME + t^0 = t_1 + t_2$ terminate the current functioning of the transition.

In the book by Atanassov (1997) we can find other versions of the above described algorithm which are based on the introduced concept of *abstract transition*. The algorithms can be considered as the most general form of functioning algorithms of generalized nets.

2.4 Algebraic Aspects of Generalized Nets

There are given two transitions

$$Z_1 = \langle L_1^1, L_2^1, t_1^1, t_2^1, r^1, M^1, \square^1 \rangle, \tag{2.3}$$

$$Z_2 = \langle L_1^2, L_2^2, t_1^2, t_2^2, r^2, M^2, \square^2 \rangle \tag{2.4}$$

with prescribed elements. For Z_1 and Z_2 we can define as follows:

$$Z_1 = Z_2$$

if and only if $(i = 1, 2, \forall i : 1 \le i \le 7)\ (pr_i Z_1 = pr_i Z_2,)$

$$Z_1 \subset Z_2$$

if and only if $(\forall i : 1 \le i \le 2)(pr_i Z_1 \subset pr_i Z_2) \& (\forall i : 3 \le i \le 4)(pr_i Z_1 = pr_i Z_2)$
$\&(\forall i : 5 \le i \le 6)(pr_i Z_1 \subset pr_i Z_2) \& (pr_7 Z_1 \subset_2 pr_7 Z_2)$

where:

$\subset_1$
is a relation of inclusion over index matrices,

if $A = [K_1, L_1, \{a_{i,j}\}]$ and $B = [K_2, L_2, \{b_{i,j}\}]$

then

$A \subset_1 B$ if and only if $(K_1 \subset K_2) \& (L_1 \subset L_2) \& (\forall i \in K_1)(\forall j \in L_1)(a_{i,j} = b_{i,j})$,

$\subset_2$
is a relation of inclusion over Boolean expressions; for two expressions A and the expression A can be obtained if and only if a part of the arguments of B is removed as well as the logical operations associated to them.

Now, let us define four operations over the transitions Z_1 and Z_2. There is the following statement which is required to hold:

if the place

$$l \in pr_1 Z_i \cap pr_2 Z_i \text{ and } l \in pr_s Z_{3-i}$$

then $l \in pr_{3-s} Z_{3-i}$ for $i, s = 1, 2$.

These operations are described as follows:

the union

the necessary conditions for such operation are as follows $t_j^1 = t_j^2$ for $i = 1, 2$ and if $l \in pr_s Z_i$ then it is not allowed that $l \in pr_{3-s} Z_{3-i}$ for $i, s = 1, 2$:

$$Z_1 \cup Z_2 = \langle L_1^1 \cup L_1^2, L_2^1 \cup L_2^2, t_1^1, t_2^1, r^1 + r^2, M^1 + M^2, \vee(\Box^1, \Box^2)\rangle,$$

the intersection

under the same conditions described above:

$$Z_1 \cap Z_2 = \langle L_1^1 \cap L_1^2, L_2^1 \cap L_2^2, t_1^1, t_2^1, r^1 \times r^2, M^1 \times M^2, \wedge(\Box^1, \Box^2)\rangle,$$

the composition

under the same conditions described above and with the additional condition $L_1^1 \cap L_1^2 = L_2^1 \cap L_2^2 = \varnothing$):

$$Z_1 \circ Z_2 = \langle L_1^1 \cup (L_1^2 - L_2^1),\ L_2^2 \cap (L_2^1 - L_1^2), t_1^1, t_2^1 + t_2^2, r^1.r^2, M^1.M^2, \vee(\Box^1, \overline{\Box}^2)\rangle,$$

where $\overline{\Box}$ can be obtained from $\Box$ after removing all its arguments whose identifiers are elements of the set $L_2^1 \cup L_1^2$.

the difference

under the same conditions described above:

$$Z_1 - Z_2 = \begin{cases} Z_\varnothing & \text{if } L_1^1 \subset L_1^2 \text{ or } L_1^2 \subset L_2^2 \\ Z_1 - Z_2 = & \langle L_1^1 - L_1^2, L_2^1 - L_2^2, t_1^1, t_2^1, r^1 - r^2, M^1 - M^2, \\ & \Box^1 / \Box^2\rangle, \text{ otherwise} \end{cases}$$

where $\Box^1 / \Box^2$ are obtained from $\Box^1$ by removing all its arguments whose identifiers are elements of the set $L_1^1 \cap L_2^1$.

The problem of algebraic structures exists in any net, for example some algebraic properties of Petri nets can be found in (Crespi-Reghizzi and Mandrioli 1976, Kotov 1978).

2.5 Operator Aspects of Generalized Nets

In Sect. 2.4 we recalled the operations and relations defined over the transitions of generalized nets, similarly there are defined operations over generalized nets.

It should be mentioned that operations like *union*, *intersection*, *composition* and *iteration* do not exist in the standard Petri net theory. These operations can be obtained for other types of Petri nets while they are simply involved in the theory of generalized nets. These operations are useful not only for constructing generalized net descriptions of real discrete processes, but also the operator aspects are important in the theory of generalized nets (Atanassov 1991, 1992, 1997, 1998, 2007).

Here we recall six types of operators defined for generalized nets. The main feature of these operators is following, namely each operator gives possibility to change a given generalized net into a new generalized net with different description as well as with new desired properties. There are the following groups of operators:

- global G-operators,
- local P-operators,
- hierarchical H-operators,
- reducing R-operators,
- extending O-operators,
- dynamic D-operators.

Global operators transform the whole given generalized net (or all elements) to a given type according to a definite procedure. There are the operators to change:

- the form and structure of the transitions (G_1, G_2, G_3, G_4, G_6),
- the temporal elements of the considered generalized net (G_7, G_8),
- the duration of its functioning (G_9),
- the set of tokens (G_{10}),
- the set of the initial characteristics (G_{11}),
- the characteristic function of the net (G_{12}),
- the evaluation function (G_{13}),
- other net's functions (G_5, G_{14}, G_{15}, G_{16}, G_{17}, G_{18}, G_{19}, G_{20}).

Exemplary functioning of the global operators following:

- G_2-operator can divide a given generalized net a single transition,
- G_4-operator gives the possibilities to add two special places: the first one is responsible for additional general input place in which all tokens enter the generalized net and then the tokens are spread to the real input places and second one is responsible for additional general output place which collects all tokens passing the generalized net,
- G_3-operator reduces a given generalized net, taking into account all places and all tokens which did not take place within the functioning process, and, as a result, the new generalized net has different description but the same functioning,
- other global operators: G_5-operator, G_{12}-operator, G_{13}-operator, G_{14}-operator, G_{15}-operator, G_{16}-operator, G_{17}-operator, G_{18}-operator, G_{19}-operator, G_{20}-operator are responsible for changing other different functions of the generalized net.

The *local operators* are responsible for transforming single elements of transitions of a considered generalized net. We can distinguish three types of local operators:

- o P_1-operator, P_2-operator, P_3-operator, P_4-operator are temporal and allow to change the temporal elements of a considered transition,
- o P_5-operator, P_6-operator have a matrix form and can be used to change some of the index matrices of a considered transition,
- o P_7-operator changes the transition's type,
- o P_8-operator changes the capacity of some of the places,
- o P_9-operator changes the characteristic function of an output place,
- o P_{10}-operator changes the function associated with the transition condition predicates of the transition.

The hierarchical operators consist of five different types and can be grouped into two categories according to their way of functioning:

- o H_1-operator, H_3–operator, H_6-operator, H_7-operator and partially H_5 allow to expand a generalized net,
- o H_2-operator, H_4-operator and partially H_5-operator allow shrinking a generalized net.

In the case of H_5-operator it can be said that it has both properties - expanding as well as shrinking.

Considering the hierarchical operators from object of action point of view we can distinguish operators:

- o H_1-operator and H_2-operator giving as a result a new place,
- o H_3-operator, H_4-operator and H_5-operator giving as a result a new transition.

The hierarchical operator H_1-operator replaces a place by a new place or even a new generalized net, while H_3-operator replaces a transition by a new transition or a new generalized net. Conversely, H_2-operator and H_4-operator replace a part of a generalized net by a single place or a single transition, respectively, while H_5-operator allows changing a subnet of a generalized net by another subnet.

Summarizing, the expanding operators can be applied to enlarge the generalized net description of the process, while the shrinking operators allow to aggregate details of the modelled process.

The reducing operators serve to develop a new reduced generalized net description, meanwhile the extending operators allow obtaining any extensions of generalized net.

There are also dynamic operators which are responsible for the functioning of generalized nets, and can be shortly explained as follows:

- o operators $D(1,i)$, $i = 1, 2, ..., 18$, determine the procedure of evaluating the transition condition predicates,
- o operators $D(2, 1)$, $D(2, 2)$, $D(2, 4)$, $D(2, 3)$, are responsible for splitting of tokens,

- operators $D(3, 1)$ and $D(3, 2)$ determine the ways of the token transfer via the transition,
- operators $D(4, 1)$, $D(4, 2)$, $D(4, 3)$ and $D(4, 4)$ determine the ways of evaluating the transition condition predicates.

Details of generalized net operators can be found in many papers and books written first of all by Krassimir Atanassov. Most of the works have been developed in a very theoretical way. It seems that they are important not only from the theoretical point of view, but also for practical applications. The operators allow users to investigate functioning of discrete event systems.

2.6 Concept of Index Matrices

In the description of the transitions in the generalized nets the *index matrices* (Atanassov 1987, 1997) are used, which will be introduced here in brief.

Let I be a fixed set of indices and R be the set of real numbers. By an index matrix (IM) with index sets K and L $(K, L \subset I)$ we will mean the following object:

$$[K, L, \{a_{k_i,l_j}\}] \equiv \begin{array}{c|cccc} & l_1 & l_2 & \cdots & l_n \\ \hline k_1 & a_{k_1,l_1} & a_{k_1,l_2} & \cdots & a_{k_1,l_n} \\ k_2 & a_{k_2,l_1} & a_{k_2,l_2} & \cdots & a_{k_2,l_n} \\ \vdots & & & & \\ k_m & a_{k_m,l_1} & a_{k_m,l_2} & \cdots & a_{k_m,l_n} \end{array}$$

where $K = \{k_1, k_2, \ldots, k_m\}$, $L = \{l_1, l_2, \ldots, l_n\}$, $a_{k_i,l_j} \in R$ for $i = 1, 2, \ldots, m$, $j = 1, 2, \ldots, n$.

The usual matrix operations like *addition* and *multiplication* are also defined for the index matrices $A = [K, L, \{a_{k_i,l_j}\}]$ and $B = [P, Q, \{b_{p_r,q_s}\}]$ in the following way:

$$A + B = [K \cup P, L \cup Q, \{c_{t_u,v_w}\}]$$

where

$$c_{t_u,v_w} = \begin{cases} a_{k_i,l_j} & \text{if } t_u = k_i \in K \text{ and } v_w = l_j \in L - Q \\ & \text{or } t_u = k_i \in K - P \text{ and } v_w = l_j \in L \\ b_{p_r,q_s} & \text{if } t_u = p_r \in P \text{ and } v_w = q_s \in Q - L \\ & \text{or } t_u = p_r \in P - K \text{ and } v_w = q_s \in Q \\ a_{k_i,l_j} + b_{p_r,q_s} & \text{if } t_u = k_i = p_r \in K \cap P \\ & \text{and } v_w = l_j = q_s \in L \cap Q \\ 0 & \text{otherwise} \end{cases}$$

$A \times B = [K \cap P, L \cap Q, \{c_{t_u,v_w}\}]$,

where

$$c_{t_u,v_w} = a_{k_i,l_j} \cdot b_{p_r,q_s},$$

for $t_u = k_i = p_r \in K \cap P$ and $v_w = l_j = q_s \in L \cap Q$.

$A \cdot B = [K \cup (P - L), Q \cup (L - P), \{c_{t_u,v_w}\}]$,

where

$$c_{t_u,v_w} = \begin{cases} a_{k_i,l_j} & \text{if } t_u = k_i \in K \text{ and } v_w = l_j \in L - P \\ b_{p_r,q_s} & \text{if } t_u = p_r \in P - L \text{ and } v_w = q_s \in Q \\ \sum\limits_{l_j = p_r \in L \cap P} a_{k_i,l_j} b_{p_r,q_s} & \text{if } t_u = k_i \in K \text{ and } v_w = q_s \in Q \\ 0 & \text{otherwise} \end{cases}$$

$A - B = [K - P,\ L - Q,\ \{c_{t_u,v_w}\}]$,

where "–" is the set theoretic difference operation and

$$c_{t_u,v_w} = a_{k_i,l_j},.$$

for $t_u = k_i \in K - P$ and $v_w = l_j \in L - Q$.

2.7 Research on Generalized Nets

In the last thirty years lots of publications related to the generalized net theory as well as applications of generalized nets have been published. Here, we will present a list of categories and topics undertaken within the generalized nets research.

The first category of research is devoted to the theory of generalized nets and the following topics can be distinguished:

- the definitions introduced and developed by Krassimir Atanassov in many papers, e.g. (Atanassov 1984, 1987, 1991, 2007),
- the basic properties, e.g. (Atanassov 1984, 1987, 1991, 2007),
- the reduced generalized nets, e.g. (Atanassov 1984, 1987, 1991, 2007),
- the extended generalized nets, e.g. (Atanassov 1984, 1987, 1991, 2007),
- the intuitionistic fuzzy generalized nets; the first paper merging the Petri net type with fuzzy sets was written by Atanassov (1985) and next e.g. (Atanassov 1991, 2007),
- the algebraic aspects of generalized nets, e.g. Atanassov (1991, 2007), Atanassov and Krawczak (2005), Krawczak (2006b),
- the topological aspects of generalized net theory, e.g. Atanassov (1991, 2007), Dincheva and Atanassov (1990),
- the logical aspects of generalized nets, e.g. (Atanassov 1991),
- the methological aspects of generalized nets, e.g. (Atanassov 1991, 2007),
- the tools for analyzing the results of the functioning of generalized nets and the functioning of different types of Petri nets, e.g. (Atanassov 1991).

The second category is devoted to relations of generalized nets to other objects:

- finite automata and recursive functions and Turing machine, e.g. (Atanassov 1991, Krawczak 2003a),
- system theory, e.g. (Atanassov 1992, 1997),
- modelling of abstract processes, e.g. Atanassov (1997), Atanassov and Aladjov (2000),
- artificial intelligence, e.g. Atanassov (1994, 1998), Chountas et al. (2007), Kolev et al. (2006), Krawczak et al. (2008),
- data bases and expert systems and data mining, e.g. (Atanassov 1998, Sotirova 2006),
- machine learning, e.g. Atanassov and Aladjov (2000), Chountas et al. (2007), Shannon et al. (2005), Shannon et al. (2007),
- the generalized nets and flexible manufacturing systems and robotics, e.g. (Atanassov 1991, 2007),
- pattern and speech recognition, e.g. (Atanassov et al. 2006),
- decision making, planning and optimization, e.g. Melo-Pinto et al. (2005), Shannon et al. (2007),
- neural networks, e.g. Hadjyisky and Atanassov (1995), Kuncheva and Atanassov (1996), Krawczak (2003e), Sotirov (2006), Atanassov et al. (2009), Krawczak et al. (2010), Sotirov and Krawczak (2011, 2012), Krawczak et al. (2012),

In the third category the generalized nets are applied in:

- medicine, e.g. (Atanassov 1993, Choy et al. 2007),
- biotechnology, e.g. Shannon et al. (2004),
- astronomy, e.g. (Atanassov 1993),
- ecology, e.g. (Choy, Krawczak, Shannon and Szmidt 2007),
- computer science, e.g. (Chountas et al. 2007, Koneva and Atanassov 1990a, Krawczak et al. 2007),
- transport systems, e.g. (Koneva and Atanassov 1990b, Atanassov 1991),
- industry, e.g. Dimitrova et al. (1993),
- and others.

The above mentioned topics can be also found in books where there are many excellent chapters related to the theory and applications of generalized nets, exemplary books edited among other by Atanassov, Kacprzyk, Krawczak and Szmidt are included in References: [17], [18], [19], [22], [23], [24], [25], [26], [27], [28], [29], [30], [31], [32], [33], [34], [35] and [95].

Chapter 3
Simulation Process of Neural Networks

3.1 Multistage Neural System

Due to Cybenko's theorem, described in Sect. 1.5, multilayer feedforward neural networks are universal approximators or universal classifiers. Approximation or classification done by neural networks is performed during the simulation process described in Sect. 1.2.

According to the notation used in Chap. 1 a multilayer neural network consists of $L+1$ layers, each layer labelled by $l = 0,1,2,...,L$, the layer $l = 0$ denoting the external inputs to the network. Each layer is composed of $N(l)$ neurons, $l = 0,1,2,...,L$, where $N(0)$ denotes the number of inputs. The neurons belonging to the $(l-1)$-st layer are connected with the neurons of the l-th layer.

The simulation process of neural networks can be expressed by *generalized nets* methodology. There are several scientific works developing neural networks simulation process in the setting of generalized net methodology. The first papers by Hadjyisky and Atanassov (1990a, 1990b, 1993) described the basic elements of neural networks by generalized nets, and others were mostly written by Krawczak (2003d, 2003e, 2004a, 2004f, 2005d), and by Krawczak and Aladjov (2003), as well as in books by Krawczak (2003b, 2003c). In this chapter, we will summarize those results. At the same time, this summarizing will be the beginning of introducing generalized net theory to describe feedforward neural network simulation process in a different way – a counterpart way.

Additionally, we will consider multilayer feedforward neural networks as multistage systems; such approach was introduced in several papers by Krawczak and Mizukami (1994) and by Krawczak (1997, 2001a, 2002a, 2002b, 2004d).

In this chapter we will consider the multilayer neural network as a system (Bryson and Ho 1969, Bellman 1972, Mesarovic and Takahara 1975, Morrison 1996). The system consists of a set of connected subsystems (Krawczak 2003b, 2003c, 2003d, 2003e, 2004b, 2004c, 2004f, 2005a, 2005c). Fig. 3.1 shows a multilayer neural network as a system with subsystems, with the way of distinguishing the subsystems called aggregation.

M. Krawczak: *Multilayer Neural Networks*, SCI 478, pp. 31–69.
DOI: 10.1007/978-3-319-00248-4_3

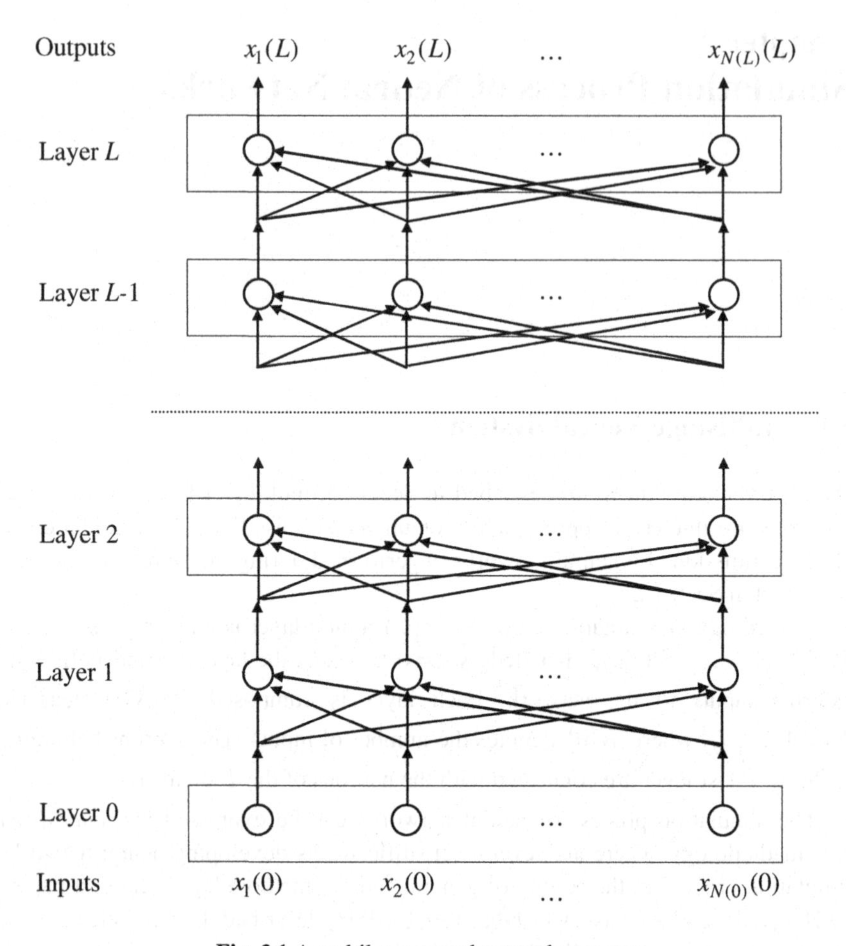

Fig. 3.1 A multilayer neural network structure

We will consider three main cases of aggregation of subsystems within any multilayer neural network, namely:

A. there is no aggregation – i.e. any neuron is treated as a subsystem,
B. the neurons are aggregated within each layer – the separate layers are the subsystems,
C. full aggregation –neither separate neurons nor layers are distinguished.

In the case A (Krawczak 2003b, 2003c), the considered neural network consists of the following number of subsystems (neurons):

$$NL = \sum_{l=1}^{L} N(l) \tag{3.1}$$

described by the activation function as follows

$$x_{pj(l)} = f\left(net_{pj(l)}\right), \tag{3.2}$$

where

$$net_{pj(l)} = \sum_{i(l-1)=1}^{N(l-1)} w_{i(l-1)j(l)}\, x_{pi(l-1)}, \tag{3.3}$$

while $x_{pi(l-1)}$ denotes the output of the i-th neuron with respect to the pattern p, $p = 1,2,\ldots,P$, and the weight $w_{i(l-1)j(l)}$ connects the $i(l-1)$-th neuron from the $(l-1)$-st layer with the $j(l)$-th from the l-th layer, $j(l) = 1,2,\ldots,N(l)$, $l = 1,2,\ldots,L$.

In the case B (Krawczak 2003b, 2003c), there are L subsystems with aggregated neurons within each layer, and each subsystem (layer) is described as follows:

$$X(l) = F_{(l)}(W(l-1), X(l-1)) \quad \text{for} \quad l = 1,2,\ldots,L \tag{3.4}$$

where $X(l)$ denotes the aggregated output of the layer l

$$X(l) = \left[x_{1(l)}, x_{2(l)},\ldots, x_{N(l)}\right]^T \quad \text{for} \quad l = 0,1,2,\ldots,L, \tag{3.5}$$

while $W(l-1)$ denotes the aggregated weights connecting the l-th layer with the $(l-1)$-st layer

$$W(l-1) = \left[w_{1(l-1)}, w_{2(l-1)},\ldots, w_{N(l-1)}\right]^T \tag{3.6}$$

and

$$w_{i(l-1)} = \left[w_{i(l-1)1(l)}, w_{i(l-1)2(l)},\ldots, w_{i(l-1)N(l)}\right]^T. \tag{3.7}$$

The weight $w_{i(l-1)j(l)}$ connects the $i(l-1)$-th neuron from the $(l-1)$-st layer with the $j(l)$-th from the l-th layer, and $X(l-1)$ is the aggregated output of the $(l-1)$-st layer, $j(l) = 1,2,\ldots,N(l)$, $l = 1,2,\ldots,L$.

In the case C (Krawczak 2003b, 2003c) there are no subsystems distinguished within the network, and the whole network is treated as a system and is represented by

$$X(L) = G(WW(L), X(0)) \tag{3.8}$$

where

$$WW(L) = [W(1), W(2),\ldots, W(L)]^T \tag{3.9}$$

$$W(l)=\left[w_{1(l)}, w_{2(l)},..., w_{N(l)}\right]^T \tag{3.10}$$

$$w_{i(l-1)}=\left[w_{i(l-1)1(l)}, w_{i(l-1)2(l)},..., w_{i(l-1)N(l)}\right]^T,$$

while $w_{i(l-1)j(l)}$ is the weight connecting the $i(l-1)$-st neuron from the $(l-1)$-st layer with the $j(l)$-th one from the l-th layer, $l=1,2,...,L$, $j(l)=1,2,...,N(l)$.

It is obvious that the different cases of aggregation of the neural network determine different streams of information passing through the system. In the subsequent sections we will describe the way of modelling the simulation process of multilayer neural networks by generalized nets for these three cases of aggregation.

3.2 Case A – Without Aggregation

Let us consider the general structure of multilayer neural networks shown in Fig. 3.1. One of the possible generalized net representations of the simulation process of this neural network is depicted in Fig. 3.2.

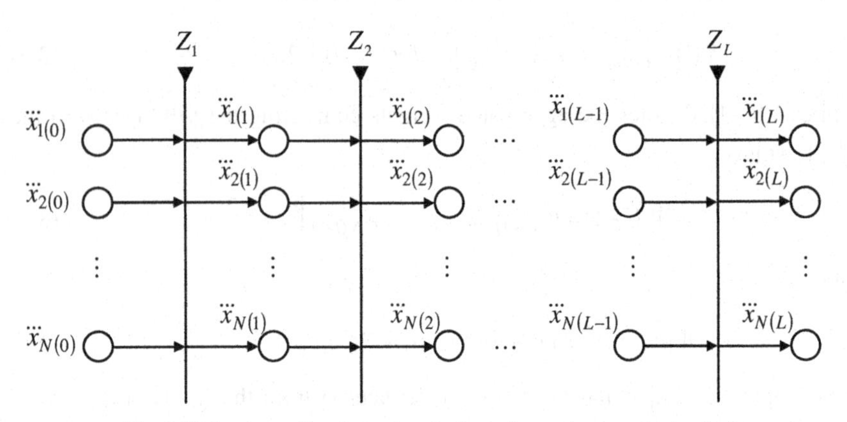

Fig. 3.2 The generalized net description of neural network simulation

The generalized net from Fig. 3.2, modelling the simulation process of the multilayer neural network from Fig. 3.1, consists of a set of L transitions, each transition being of the following form

$$Z_l=\left\langle\left\{\dddot{x}_{1(l-1)}, \dddot{x}_{2(l-1)},..., \dddot{x}_{N(l-1)}\right\}, \left\{\dddot{x}_{1(l)}, \dddot{x}_{2(l)},..., \dddot{x}_{N(l)}\right\}, \tau_l, \tau'_l, r_l, M_l, \square_l\right\rangle \tag{3.11}$$

for $l=1,2,...,L$,

where

$\left\{\dddot{x}_{1(l-1)}, \dddot{x}_{2(l-1)}, \ldots, \dddot{x}_{N(l-1)}\right\}$

is the set of input places of the l -th transition,

$\left\{\dddot{x}_{1(l)}, \dddot{x}_{2(l)}, \ldots, \dddot{x}_{N(l)}\right\}$

is the set of output places of the l -th transition,

τ_l

is the time when the l -th transition is fired out, while it is assumed that

$$\tau_1 = T \text{ and } \tau_l = T + \sum_{k=2}^{l} \tau'_{k-1},$$

τ'_l

is the duration time of firing of the l -th transition,

r_l

denotes the l -th transition condition determining the transfer of tokens from the transition's inputs $\left\{\dddot{x}_{1(l-1)}, \dddot{x}_{2(l-1)}, \ldots, \dddot{x}_{N(l-1)}\right\}$ to its outputs $\left\{\dddot{x}_{1(l)}, \dddot{x}_{2(l)}, \ldots, \dddot{x}_{N(l)}\right\}$, and has the following index matrix form:

$$r_l = \begin{array}{c|cccc} & \dddot{x}_{1(l)} & \dddot{x}_{2(l)} & \cdots & \dddot{x}_{N(l)} \\ \hline \dddot{x}_{1(l-1)} & true & true & \cdots & true \\ \dddot{x}_{2(l-1)} & true & true & \cdots & true \\ \vdots & \vdots & \cdots & \cdots & \vdots \\ \dddot{x}_{N(l-1)} & true & true & \cdots & true \end{array} \tag{3.12}$$

where the value *true* indicates that the tokens representing the neurons can be transferred from the $i(l-1)$-st input place to the $j(l)$-th output place, $i(l-1) = 1, 2, \ldots, N(l-1)$, $j(l) = 1, 2, \ldots, N(l)$,

M_l

indicates an index matrix describing the capacities of transition's arcs:

$$M_l = \begin{array}{c|cccc} & \dddot{x}_{1(l)} & \dddot{x}_{2(l)} & \cdots & \dddot{x}_{N(l)} \\ \hline \dddot{x}_{1(l-1)} & 1 & 1 & \cdots & 1 \\ \dddot{x}_{2(l-1)} & 1 & 1 & \cdots & 1 \\ \vdots & \vdots & \cdots & \cdots & \vdots \\ \dddot{x}_{N(l-1)} & 1 & 1 & \cdots & 1 \end{array} \tag{3.13}$$

$\square_l$

has a form of Boolean expression $\wedge\left(\dddot{x}_{1(l-1)}, \dddot{x}_{2(l-1)},..., \dddot{x}_{N(l-1)}\right)$ and stipulates that each input place $\dddot{x}_{i(l-1)}$, $i(l-1)=1,2,..., N(l-1)$, must contain a token that will be transferred to the l -th transition.

The generalized net describing the considered neural network simulation process (Fig. 2.3) has the following form:

$$GN1 = \left\langle \left\langle A, \pi_A, \pi_X, c, g, \Theta_1, \Theta_2 \right\rangle, \left\langle K, \pi_k, \Theta_K \right\rangle, \left\langle T, t^0, t^* \right\rangle, \left\langle Y, \Phi, b \right\rangle \right\rangle \tag{3.14}$$

The first part of generalized net (3.14) $\left\langle A, \pi_A, \pi_X, c, g, \Theta_1, \Theta_2 \right\rangle$ consists of the following parameters:

$A = \{Z_1, Z_2,..., Z_L\}$
is the set of transitions,

π_A
is a function classifying the transitions, this classification giving the priorities of the transitions, i.e. $\pi_A : A \rightarrow N$, where $N = \{0, 1, 2, \ldots\} \cup \{\infty\}$ - in the considered neural network case this function is not valid because the transitions are arranged in a natural way (the asterisk * will be used in the subsequent text in order to denote the elements of the general net structure which can be omitted),

π_X
is a function describing the priorities of the places in the following way:

$$pr_1\{Z_1, Z_2,..., Z_L\} = \{\dddot{x}_{1(0)}, \dddot{x}_{2(0)},..., \dddot{x}_{N(0)}, \dddot{x}_{1(1)}, \dddot{x}_{2(1)},..., \dddot{x}_{N(1)},..., \dddot{x}_{1(L-1)}, \dddot{x}_{2(L-1)},..., \dddot{x}_{N(L-1)}\} \tag{3.15}$$

$$pr_2\{Z_1, Z_2,..., Z_L\} = \{\dddot{x}_{1(1)}, \dddot{x}_{2(1)},..., \dddot{x}_{N(1)}, \dddot{x}_{1(2)}, \dddot{x}_{2(2)},..., \dddot{x}_{N(2)},..., \dddot{x}_{1(L)}, \dddot{x}_{2(L)},..., \dddot{x}_{N(L)}\} \tag{3.16}$$

$$pr_1 A \cup pr_2 A = \{\dddot{x}_{1(0)}, \dddot{x}_{2(0)},..., \dddot{x}_{N(0)}, \dddot{x}_{1(1)}, \dddot{x}_{2(1)},..., \dddot{x}_{N(1)}, \dddot{x}_{1(2)}, \dddot{x}_{2(2)},..., \dddot{x}_{N(2)},... ..., \dddot{x}_{1(L)}, \dddot{x}_{2(L)},..., \dddot{x}_{N(L)}\} \tag{3.17}$$

c

is a function describing the capacities of the places; in our case the capacity function has the form:

$$c\left(x_{i(l)}\right)=1\ ,\quad \text{for } i(l)=1,2,\ldots,N(l),\ l=0,1,2,\ldots,L\ , \tag{3.18}$$

g

is a function that calculates the truth values of the predicates of the transition conditions (3.12); for the generalized net described here this function is constant, i.e.

$$g\left(r_{l,i(l-1)j(l)}\right)=true \tag{3.19}$$

(instead of the value *false* or *true* we can use the values 0 or 1),

Θ_1

is a function yielding the next time-moment when the transitions can be again activated, and we can consider two cases:

a) performing the simulation of the neural network for only one input

$$X(0)=\left[x_{1(0)},\ x_{2(0)},\ldots,\ x_{N(0)}\right]^T \tag{3.20}$$

and it means that each transition fires out only once, and there is no next time the transition is fired, this case denoted *,

b) performing the simulation of the neural network for a set of inputs

$$\{X_1(0),X_2(0),\ldots,X_P(0)\} \tag{3.21}$$

where P denotes the number of samples (for example, when we use the neural network as an approximator of some function and we have P samples), and in this case each transition will be active P times; the function Θ_1 describing the time of a new firing for each transition has the following form

$$\Theta_1(t_l)=t_l',\ l=1,2,\ldots,L \tag{3.22}$$

where (assuming that $pr_3 Z_0=T$)

$$\begin{aligned} t_l &= T+(p-1)\sum_{k=1}^{L} pr_4 Z_k \\ &\quad + pr_3 Z_{l-1} = T+(p-1)+\sum_{k=1}^{L}\tau_k' + pr_3 Z_{l-1} \end{aligned} \tag{3.23}$$

and

$$t_l' = t_l + \sum_{k=1}^{L} pr_4 Z_k = t_l + \sum_{k=1}^{L} \tau_k' \tag{3.24}$$

$t' \in [T, T + t^*]$ and $t \le t'$ for $p = 1, 2, ..., P$; the value of this function is calculated at the moment when the transition terminates being active,

Θ_2

is a function giving the duration of activity of a given transition Z_l

$$\Theta_2(t_l) = t_l'', \; l = 1, 2, ..., L \tag{3.25}$$

where t_l is described by (3.23) and

$$t_l'' = pr_4 Z_l = \tau_l' \tag{3.26}$$

the value of this function is calculated at the moment when the transition starts functioning.

The second part of generalized net (3.12) $\langle K, \pi_k, \Theta_K \rangle$ consists of the following parameters:

K

is the set of tokens entering the generalized net, in the considered case there are $N(0)$ input places and each place contains one token; this set can be written as

$$K = \{\alpha_{1(0)}, \alpha_{2(0)}, ..., \alpha_{N(0)}\} \tag{3.27}$$

π_K

is a function describing the priorities of the tokens, here all tokens have the same priorities, and it will be denoted by "*" for $\pi_K(\alpha_{l(0)})$ and $l(0) = 1, 2, ..., N(0)$,

Θ_K

is a function giving the time-moment when a given token can enter the net, i.e. all the tokens enter the considered generalized net at the same moment T,

The third part of generalized net (3.14) $\langle T, t^0, t^* \rangle$ cinsists of the following parameters:

T

is the time when the generalized net starts functioning – here it is assumed that the net starts at the moment T, when the tokens enter the net,

t^0

is an elementary time-step, here this parameter is not used and is denoted by $*$,

t^*

determines the duration of the generalized net functioning, that is:

a) for performing the simulation for only one input

$$t^* = \sum_{l=1}^{L} \tau'_l \tag{3.28}$$

b) for performing the simulation for a set of inputs

$$t^* = P \sum_{l=1}^{L} \tau'_l \tag{3.29}$$

where P denotes the number of samples,

The forth part of generalized net (3.14) $\langle Y, \Phi, b \rangle$ cinsists of the following parameters:

Y

denotes the set of all the initial characteristics of the tokens, the characteristics of tokens describe the information which is carried by tokens and changed in transitions,

$$Y = \{ y(\alpha_{1(0)}), y(\alpha_{2(0)}), ..., y(\alpha_{N(0)}) \} \tag{3.30}$$

where

$$y(\alpha_{i(0)}) = \langle NN1, N(0), N(1), imX_{i(0)}, imW_{i(0)}, F_{(1)}, imout_{i(0)} \rangle \tag{3.31}$$

is the initial characteristic of the token $\alpha_{i(0)}$ that enters the place $\ddot{x}_{i(0)}$, $i(0) = 1, 2, ..., N(0)$, where

$NN1$
the neural network identifier,

$N(0)$
number of input places to the net as well as to the transition Z_1 (equal to the number of inputs to the neural network),

$N(1)$
number of the output places of the transition Z_1,

$$imX_{i(0)} = [0, ..., 0, x_{i(0)}, 0, ..., 0]^T \tag{3.32}$$

is the index matrix, indicating the inputs to the network, of dimension $N(0) \times 1$ in which all elements are equal 0 except for the element $i(0)$ whose value is equal $x_{i(0)}$ (the $i(0)$-th input of the neural network),

$$imW_{i(0)} = \begin{array}{c|c|c|c|c} & \dddot{x}_{1(1)} & \dddot{x}_{2(1)} & \cdots & \dddot{x}_{N(1)} \\ \hline \dddot{x}_{i(0)} & w_{i(0)1(1)} & w_{i(0)2(1)} & \cdots & w_{i(0)N(1)} \end{array} \quad (3.33)$$

has a form of an index matrix and denotes the weights connecting the $i(0)$-th input with all neurons allocated to the 1-st layer,

$$F_{(1)} = \left[f_{1(1)}\left(\sum_{i(0)=1}^{N(0)} x_{i(0)} w_{i(0)1(1)} \right), f_{2(1)}\left(\sum_{i(0)=1}^{N(0)} x_{i(0)} w_{i(0)2(1)} \right), \ldots, f_{N(1)}\left(\sum_{i(0)=1}^{N(0)} x_{i(0)} w_{i(0)N(1)} \right) \right]^T \quad (3.34)$$

denotes a vector of the activation functions of the neurons associated with the 1-st layer,

$$imout_{i(0)} = \begin{array}{c|c|c|c|c} & \dddot{x}_{1(1)} & \dddot{x}_{2(1)} & \cdots & \dddot{x}_{N(1)} \\ \hline \dddot{x}_{i(0)} & x_{i(0)} w_{i(0)1(1)} & x_{i(0)} w_{i(0)2(1)} & \cdots & x_{i(0)} w_{i(0)N(1)} \end{array} \quad (3.35)$$

is an index matrix describing the signal outgoing from the $i(0)$-th input place, $i(0) = 1, 2, \ldots, N(0)$, to all output places of the Z_1 transition,

Φ

is a characteristic function that generates the new characteristics of the new tokens after passing the transition; for the transition Z_l, $l = 1, 2, \ldots, L$, there are $N(l-1)$ input places $\{\dddot{x}_{1(l-1)}, \dddot{x}_{2(l-1)}, \ldots, \dddot{x}_{N(l-1)}\}$ and with each place there is associated a single token $\alpha_{i(l-1)}$, $i(l-1) = 1, 2, \ldots, N(l-1)$, having the characteristic

$$y(\alpha_{i(l-1)}) = \langle NN1, N(l-1), N(l), imX_{i(l-1)}, imW_{i(l-1)}, F_{(l)}, imout_{i(l-1)} \rangle \quad (3.36)$$

where

$NN1$

the neural network identifier,

$N(l-1)$

number of input places to the net as well as to the transition Z_l,

$N(l)$

number of the output places of the transition,

$$imX_{i(l-1)} = [0, \ldots, 0, x_{i(l-1)}, 0, \ldots, 0]^T \quad (3.37)$$

is the index matrix of dimension $N(l-1)\times 1$ in which all elements are equal 0 except the element $i(l-1)$ whose value is equal $x_{i(l-1)}$ - the $i(l-1)$-st input value associated with the Z_l transition,

$$imW_{i(l-1)} = \begin{array}{c|c|c|c|c} & \dddot{x}_{1(1)} & \dddot{x}_{2(1)} & \cdots & \dddot{x}_{N(1)} \\ \hline \dddot{x}_{i(l-1)} & w_{i(l-1)1(l)} & w_{i(l-1)2(l)} & \cdots & w_{i(l-1)N(l)} \end{array} \quad (3.38)$$

is an index matrix describing the weight connection between the $i(l-1)$-st input places with all output places of the Z_l transition,

$$F_{(l)} = \left[f_{1(l)}\left(\sum_{i(l-1)=1}^{N(l-1)} x_{i(l)}\, w_{i(l-1)1(l)} \right), f_{2(l)}\left(\sum_{i(l-1)=1}^{N(l-1)} x_{i(l-1)}\, w_{i(l-1)2(l)} \right), \right.$$
$$\left. \ldots, f_{N(l)}\left(\sum_{i(l-1)=1}^{N(l-1)} x_{i(l-1)}\, w_{i(l-1)N(l)} \right) \right]^T \quad (3.39)$$

is a vector of the activation functions of the neurons associated with the l-th layer of the neural network,

$$imout_{i(l-1)} = \begin{array}{c|c|c|c|c} & \dddot{x}_{1(l)} & \dddot{x}_{2(l)} & \cdots & \dddot{x}_{N(l)} \\ \hline \dddot{x}_{i(l-1)} & x_{i(l-1)}\, w_{i(l-1)1(l)} & x_{i(l-1)}\, w_{i(l-1)2(l)} & \cdots & x_{i(l-1)}\, w_{i(l-1)N(l)} \end{array} \quad (3.40)$$

is an index matrix describing the signals outgoing from the $i(l)$-th input place, $i(l)=1,2,\ldots,N(l)$, to all output places of the Z_l transition.

The tokens $\alpha_{i(l-1)}$, $i(l-1)=1,2,\ldots,N(l-1)$, passing the transition Z_l vanish, and the new tokens $\alpha_{j(l)}$, $j(l)=1,2,\ldots,N(l)$, associated with the output places $\{\dddot{x}_{1(l)}, \dddot{x}_{2(l)},\ldots, \dddot{x}_{N(l)}\}$ of the transition Z_l are generated, their characteristics being described as follows

$$y(\alpha_{j(l)}) = \langle NN1, N(l), N(l+1), imX_{j(l)}, imW_{j(l)}, F_{(l+1)}, imout_{j(l)} \rangle \quad (3.41)$$

for $l=1,2,\ldots,L-1$, while

$$y(\alpha_{j(L)}) = \langle NN1, N(L), imX_{j(L)} \rangle, \quad (3.41a)$$

and for these new tokens the values $x_{j(l)}$, $j(l)=1,2,\ldots,N(l)$, are calculated in the following way

$$imX_{j(l)} = f_{j(l)}\left(\sum_{i(l-1)=1}^{N(l-1)} imout_{i(l-1)}\right), \quad l = 1,2,...,L. \tag{3.42}$$

It should be mentioned here that $imX_{j(L)}$, $j(L) = 1,2,...,N(L)$, denotes the output of the neural network, the final state of the network after ending the simulation process,

b
is a function describing the maximum number of characteristics a given token can receive; in the here considered neural network simulation process this function has a simple form

$$b(\alpha_{j(l)}) = 1, \text{ for } j(l) = 1,2,...,N(l), \; l = 1,2,...,L, \tag{3.43}$$

which means that the characteristic of each token $\alpha_{j(l)}$, $j(l) = 1,2,..., N(l)$, $l = 1,2,...,L$, is constructed on the base of the characteristics of all tokens $i(l-1) = 1,2,...,N(l-1)$ from the previous layer $(l-1)$, $l = 1,2,...,L$.

Due to the above considerations the transitions now can be rewritten in the following form

$$Z_l = \langle\{\ddot{x}_{1(l-1)}, \ddot{x}_{2(l-1)},..., \ddot{x}_{N(l-1)}\}, \{\ddot{x}_{1(l)}, \ddot{x}_{2(l)},..., \ddot{x}_{N(l)}\}, \tau_l, \tau_l', *, *, \Box_l\rangle \tag{3.44}$$

for $l = 1,2,...,L$, and other elements of (3.44) are not changed.

The *reduced form* of the generalized net describing the simulation process of the neural network has the following form:

$$GN1 = \langle \langle A, *, \pi_X, c, *, \Theta_1, \Theta_2\rangle, \langle K, *, \Theta_K\rangle, \langle T, *, t*\rangle, \langle Y, \Phi, b\rangle \rangle \tag{3.45}$$

where

$A = \{Z_1, Z_2,..., Z_L\}$
is a set of transitions,

π_X
is a function describing the priorities of the places,

c
is a function describing the capacities of the places, i.e. $c(x_{i(l)}) = 1$, $i(l) = 1, 2,..., N(l)$, $l = 0,1,2,...,L$,

Θ_1

is a function yielding the next time-moment when the transitions can be again activated, $\Theta_1(t_l)=t'_l$, $l=1,2,...,L$, described by (3.23) – (3.24),

Θ_2

is a function giving the duration of activity of the transition Z_l $\Theta_2(t_l)=t''_l$, $l=1,2,...,L$, and is described by (3.23) and (3.26),

$K=\{\alpha_{1(0)},\alpha_{2(0)},...,\alpha_{N(0)}\}$

is the set of tokens entering the generalized net,

$\Theta_K=T$

for all tokens entering the net and at this moment the net starts to function,

t^*

determines the duration of the generalized net's functioning and is described by (3.28) or (3.29),

Y

denotes the set of all initial characteristics of the tokens described by

$$Y=\{y(\alpha_{1(0)}),y(\alpha_{2(0)}),...,y(\alpha_{N(0)})\}, \text{ where}$$

$$y(\alpha_{i(0)})=\langle NN1, N(0), N(1), x_{i(0)}, W_{i(0)}, F_{(1)}, out_{i(0)}\rangle, i(0)=1,2,...,N(0),$$

Φ

is a characteristic function that generates the new characteristics of the new tokens after passing the transition, and is described by (3.31), (3.36) and (3.41),

$$b(\alpha_{j(l)})=1, \text{ for } j(l)=1,2,...,N(l), l=1,2,...,L,$$

is a function describing the number of characteristics memorized by each token.

Such generalized nets with some elements missing (the elements not being valid) are called *reduced generalized nets* (Atanassov 1991, 1992, 2007). In the above version of the generalized nets representation of the simulation process of multilayer neural network we preserve the parallelism of computation.

3.3 Case B – Aggregation within Layers

Let us again consider the structure of the multilayer neural network shown in Fig. 3.1, and aggregate the neurons allocated within each layer l, $l=1,2,...,L$. In this way we can obtain the aggregated system that consists of L subsystems (Krawczak 2003b, 2003c). The aggregated layers are described in details by (3.4) – (3.7).

Two executions of the aggregation will be considered. In the first execution all input tokens will be aggregated into only one token, while in the second execution we will not aggregate tokens but by introducing some extra places (the tokens will enter each transition sequentially).

3.3.1 Execution I

In the first execution, the generalized net representation of the neural network consists of L transitions. Each transition Z_l, $l = 1, 2, ..., L$, has the following aggregated input place

$$\dddot{X}_{(l-1)} = \left\{ \dddot{x}_{1(l-1)}, \dddot{x}_{2(l-1)}, ..., \dddot{x}_{N(l-1)} \right\} \tag{3.46}$$

and the aggregated output place

$$\dddot{X}_{(l)} = \left\{ \dddot{x}_{1(l)}, \dddot{x}_{2(l)}, ..., \dddot{x}_{N(l)} \right\}. \tag{3.47}$$

The generalized net structure of such aggregated neural network is depicted in Fig. 3.3.

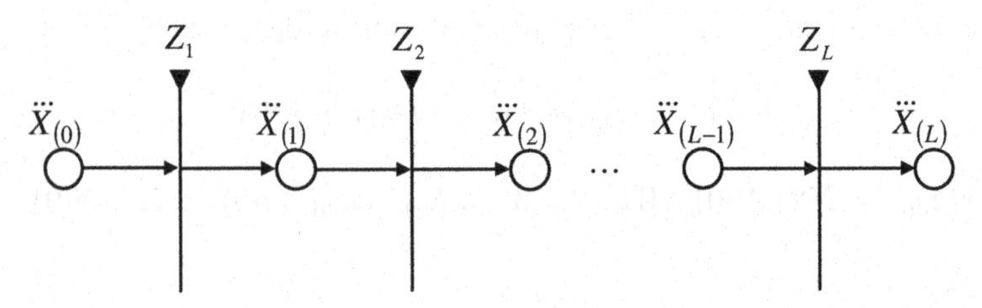

Fig. 3.3 The generalized net representation of the feedforward neural network simulation process

It is assumed that for each transition Z_l, $l = 1, 2, ..., L$, there is a single input place as well as a single output place, additionally let us assume that each place contains only one token.

Each transition Z_l, $l = 1, 2, ..., L$, has the following form

$$Z_l = \left\langle \left\{ \dddot{X}_{(l-1)} \right\}, \left\{ \dddot{X}_{(l)} \right\}, \tau_l, \tau'_l, r_l, M_l, \Box_l \right\rangle \tag{3.48}$$

where

$\left\{ \dddot{X}_{(l-1)} \right\}$

is the input place of the l-th transition,

$\left\{ \dddot{X}_{(l)} \right\}$

is the output place of the l-th transition,

τ_l

is a time when the l -th transition is fired out, while $\tau_1 = T$, $\tau_l = T + \sum_{k=2}^{l} \tau'_{k-1}$,

τ'_l

is the duration of activity of the l -th transition,

r_l

denotes the l -th transition condition of the $\alpha_{(l-1)}$ token from the transition's input place $\{\dddot{X}_{(l-1)}\}$ to its output place $\{\dddot{X}_{(l)}\}$, and it has a simple form: $r_l = true$ (i.e. the input token can be transferred to the output place without any condition),

$M_l = 1$

indicates that only one token can be transferred by arcs in the same time,

$\square_l$

is not valid due to the existence of only one token in any place.

In this way the transitions Z_l, $l = 1, 2, ..., L$, now have the following reduced form

$$Z_l = \left\langle \{\dddot{X}_{(l-1)}\}, \{\dddot{X}_{(l)}\}, \tau_l, \tau'_l \right\rangle. \tag{3.49}$$

Under the assumption that each place contains only one token the characteristics of the tokens must be of much more complex form. Let us again consider the generalized net, which describes the neural network simulation process but now with aggregated layers. The new aggregated generalized net has the following form:

$$GN2 = \left\langle \left\langle A, \pi_A, \pi_X, c, g, \Theta_1, \Theta_2 \right\rangle, \left\langle K, \pi_k, \Theta_K \right\rangle, \left\langle T, t^0, t^* \right\rangle, \left\langle Y, \Phi, b \right\rangle \right\rangle \tag{3.50}$$

The elements of the first part of (3.50) are described as follows:

$A = \{Z_1, Z_2, ..., Z_L\}$

is a set of transitions,

π_A

is not valid here,

π_X

is a function describing the priorities of the places in the following way:

$$pr_1\{Z_1, Z_2, ..., Z_L\} = \{\dddot{X}_{(0)}, \dddot{X}_{(1)}, ..., \dddot{X}_{(L-1)}\} \tag{3.51}$$

$$pr_2\{Z_1, Z_2, ..., Z_L\} = \{\dddot{X}_{(1)}, \dddot{X}_{(2)}, ..., \dddot{X}_{(L)}\} \tag{3.52}$$

$$pr_1 A \cup pr_2 A = \left\{ \dddot{X}_{(0)}, \dddot{X}_{(1)}, \ldots, \dddot{X}_{(L-1)}, \dddot{X}_{(L)} \right\} \tag{3.53}$$

$c\left(x_{i(l)}\right) = 1$

is a function describing the capacities of the places,

$g\left(r_{l,i(l-1)j(l)}\right) = true$

is a function that calculates the truth values of the predicates of the transition conditions,

Θ_1

is a function giving the next time-moment when the transitions can be again activated; for a single simulation this time is not valid, while for the simulation for a set of inputs the function has a form $\Theta_1(t_l) = t_l'$, $l = 1,2,\ldots,L$ (with $pr_3 Z_0 = T$), where

$$t_l = T + (p-1)\sum_{k=1}^{L} pr_4 Z_k + pr_3 Z_{l-1} = T + (p-1) + \sum_{k=1}^{L} \tau_k' + pr_3 Z_{l-1} \tag{3.54}$$

and

$$t_l' = t_l + \sum_{k=1}^{L} pr_4 Z_k = t_l + \sum_{k=1}^{L} \tau_k' \tag{3.55}$$

$$t' \in [T, T + t^*] \text{ and } t \le t' \text{ for } p = 1,2,\ldots,P,$$

Θ_2

is a function giving the duration of activity of each transition Z_l

$$\Theta_2(t_l) = t_l'', \; l = 1,2,\ldots,L \tag{3.56}$$

where t_l is described by (3.54) and $t_l'' = pr_4 Z_l = \tau_l'$ (the value of this function is calculated at the moment the transition starts functioning).

In the second part of the generalized net (3.50) the elements are described as follows:

$K = \left\{ \alpha_{(0)} \right\}$

is the token entering the generalized net (here only one token enters the net),

π_K

describing the priorities of the tokens is not valid here,

$\Theta_K = T$

is the time when the token $\alpha_{(0)}$ enters the net.

The elements of the third part of (3.50) constitute:

T
the time when the generalized net starts functioning,

t^0
describing the elementary time-step (not valid here),

t^*
describing the time when the generalized net is functioning:

for a single simulation $t^* = \sum_{l=1}^{L} \tau'_l$,

for a set of inputs this time is described by $t^* = P \sum_{l=1}^{L} \tau'_l$, where P denotes the number of samples.

The elements of the last part of (3.50) are described by:

$$Y = \{y(\alpha_{(0)})\}$$

is the initial characteristic of the token which enters the place $\dddot{X}_{(0)}$, the token has the form

$$y(\alpha_{(0)}) = \langle NN1, N(0), N(1), imX_{(0)}, imW_{(0)}, F_{(1)}, imout_{(0)} \rangle \tag{3.57}$$

where

$NN1$
the neural network identifier,

$N(0)$
the dimension of the input vector $X_{(0)}$,

$N(1)$
the dimension of the output vector $X_{(1)}$,

$imX_{(0)}$
the input vector to the network is obtained as a sum of the index matrices $imX_{i(0)}$, described by (3.32), from $i(0) = 1$ to $i(0) = N(0)$, and can be written as follows

$$imX_{(0)} = \sum_{i(0)=1}^{N(0)} imX_{i(0)} = [x_{1(0)}, x_{2(0)}, \ldots, x_{N(0)},]^T \tag{3.58}$$

$imX_{(1)}$
the vector of outputs of the first layer of the neural network is constructed in a similar way, and can be written as follows

$$imX_{(1)} = \sum_{j(1)=1}^{N(1)} imX_{j(1)} = [x_{1(1)}, x_{2(1)}, \dots, x_{N(1)},]^T,$$

$imW_{(0)}$

is the index matrix of weights connecting the inputs of the network with the first layer neurons, which is created as a sum of the index matrices $imW_{i(0)}$, described by (3.38), which can be written as follows

$$imW_{(0)} = \sum_{i(0)=1}^{N(0)} imW_{i(0)} =$$

	$\dddot{x}_{1(1)}$	$\dddot{x}_{2(1)}$	$\cdots$	$\dddot{x}_{N(1)}$
$\dddot{x}_{1(0)}$	$w_{1(0)1(1)}$	$w_{1(0)2(1)}$	$\cdots$	$w_{1(0)N(1)}$
$\dddot{x}_{2(0)}$	$w_{2(0)1(1)}$	$w_{2(0)2(1)}$	$\cdots$	$w_{2(0)N(1)}$
$\vdots$	$\vdots$	$\vdots$	$\cdots$	$\vdots$
$\dddot{x}_{N(0)}$	$w_{N(0)1(1)}$	$w_{N(0)2(1)}$	$\cdots$	$w_{N(0)N(1)}$

(3.59)

$$F_{(1)} = \begin{bmatrix} f_{1(1)}\left(\sum_{i(0)=1}^{N(0)} x_{i(0)} w_{i(0)1(1)}\right), f_{2(1)}\left(\sum_{i(0)=1}^{N(0)} x_{i(0)} w_{i(0)2(1)}\right), \dots \\ \dots, f_{N(1)}\left(\sum_{i(0)=1}^{N(0)} x_{i(0)} w_{i(0)N(1)}\right) \end{bmatrix}^T \qquad (3.60)$$

denotes the vector of activation functions of the neurons associated with the 1-st layer,

$imout_{(0)}$

an index matrix describing the signals outgoing from the input places of the transition Z_1 to the output places of this transition, having the form

$$imout_{(0)} = \sum_{i(0)=1}^{N(0)} imout_{i(0)} =$$

	$\dddot{x}_{1(1)}$	$\dddot{x}_{2(1)}$	$\cdots$	$\dddot{x}_{N(1)}$
$\dddot{x}_{1(0)}$	$x_{1(0)} w_{1(0)1(1)}$	$x_{1(0)} w_{1(0)2(1)}$	$\cdots$	$x_{1(0)} w_{1(0)N(1)}$
$\vdots$	$\vdots$	$\vdots$	$\cdots$	$\vdots$
$\dddot{x}_{i(0)}$	$x_{i(0)} w_{i(0)1(1)}$	$x_{i(0)} w_{i(0)2(1)}$	$\cdots$	$x_{i(0)} w_{i(0)N(1)}$
$\vdots$	$\vdots$	$\vdots$	$\cdots$	$\vdots$
$\dddot{x}_{N(0)}$	$x_{N(0)} w_{N(0)1(1)}$	$x_{N(0)} w_{N(0)2(1)}$	$\cdots$	$x_{N(0)} w_{N(0)N(1)}$

(3.61)

Φ
is a function generating the new characteristics of the new token after passing the transition Z_l, $l = 1, 2, ..., L-1$; the input place $\dddot{X}_{(l-1)}$ has a token $\alpha_{(l-1)}$ of the following characteristic

$$y(\alpha_{(l-1)}) = \langle NN1, N(l-1), N(l), imX_{(l-1)}, imW_{(l-1)}, F_{(l)}, imout_{(l-1)} \rangle \tag{3.62}$$

the token $\alpha_{(l-1)}$ passing the transition Z_l obtains the new characteristic, which can be written as follows

$$y(\alpha_{(l)}) = \langle NN1, N(l), N(l+1), imX_{(l)}, imW_{(l)}, F_{(l+1)}, imout_{(l)} \rangle \tag{3.63}$$

and this new token is associated with the output place $\dddot{X}_{(l+1)}$; the elements of (3.62) are described as follows:

$$imX_{i(l-1)} = [0, ..., 0, x_{i(l-1)}, 0, ..., 0]^T \tag{3.64}$$

$$imW_{(l-1)} = \sum_{i(l-1)=1}^{N(l-1)} imW_{i(l-1)} = \begin{array}{c|cccc} & \dddot{x}_{1(l)} & \dddot{x}_{2(l)} & \cdots & \dddot{x}_{N(l)} \\ \hline \dddot{x}_{1(l-1)} & w_{1(l-1)1(l)} & w_{1(l-1)2(l)} & \cdots & w_{1(l-1)N(l)} \\ \dddot{x}_{2(l-1)} & w_{2(l-1)1(l)} & w_{2(l-1)2(l)} & \cdots & w_{2(l-1)N(l)} \\ \vdots & \vdots & \vdots & \cdots & \vdots \\ \dddot{x}_{N(l-1)} & w_{N(l-1)1(l)} & w_{N(l-1)2(l)} & \cdots & w_{N(l-1)N(l)} \end{array} \tag{3.65}$$

$$F_{(l)} = \begin{bmatrix} f_{1(l)}\left(\sum_{i(l-1)=1}^{N(l-1)} x_{i(l)} w_{i(l-1)1(l)}\right), f_{2(l)}\left(\sum_{i(l-1)=1}^{N(l-1)} x_{i(l-1)} w_{i(l-1)2(l)}\right), ..., \\ f_{N(l)}\left(\sum_{i(l-1)=1}^{N(l-1)} x_{i(l-1)} w_{i(l-1)N(l)}\right) \end{bmatrix}^T \tag{3.66}$$

$$imout_{(l-1)} = \sum_{i(l-1)=1}^{N(l-1)} imout_{i(l-1)} =$$

$$\begin{array}{c|cccc} & \dddot{x}_{1(l)} & \dddot{x}_{2(l)} & \cdots & \dddot{x}_{N(l)} \\ \hline \dddot{x}_{1(l-1)} & x_{1(l-1)}\, w_{1(l-1)1(l)} & x_{1(l-1)}\, w_{1(l-1)2(l)} & \cdots & x_{1(l-1)}\, w_{1(l-1)N(l)} \\ \vdots & \vdots & \vdots & \cdots & \vdots \\ \dddot{x}_{i(l-1)} & x_{i(l-1)}\, w_{i(l-1)1(l)} & x_{i(l-1)}\, w_{i(l-1)2(l)} & \cdots & x_{i(l-1)}\, w_{i(l-1)N(l)} \\ \vdots & \vdots & \vdots & \cdots & \vdots \\ \dddot{x}_{N(l-1)} & x_{N(l-1)}\, w_{N(l-1)1(l)} & x_{N(l-1)}\, w_{N(l-1)2(l)} & \cdots & x_{N(l-1)}\, w_{N(l-1)N(l)} \end{array} \quad (3.67)$$

while the elements of the vector $imX_{(l)}$ are calculated in the following way

$$imX_{j(l)} = f_{j(l)}\left(\sum_{i(l-1)=1}^{N(l-1)} imout_{i(l-1)} \right), \quad (3.68)$$

$b(\alpha_{(l)}) = 1$ for $l = 1, 2, ..., L$.

The reduced form of the generalized net, describing the simulation process of the aggregated neural network has the following form:

$$GN2 = \langle \langle A, *, \pi_X, c, *, \Theta_1, \Theta_2 \rangle, \langle K, *, * \rangle, \langle T, *, t^* \rangle, \langle Y, \Phi, 1 \rangle \rangle \quad (3.69)$$

with the elements specified in detail above.

3.3.2 Execution II

Up till now the Case B was considered under the assumption that each place contains only one token and the delivered characteristics of the tokens have a complex form. It is possible to consider a structure of the generalized net similar to that from Fig. 3.3, which each place containing a number of tokens. The number of tokens in the place $\dddot{X}_{(l-1)}$, $l = 1, 2, ..., L$, corresponds to the number of neurons associated with the l-th layer. In order to introduce a prescribed number of the tokens into each place we need to change the parallelism of signal flows by the sequential flows. The generalized net representation shown in Fig. 3.3 will become a little bit more complex through the introduction of an extra place to each transition.

Let us again consider the structure of the multilayer neural network shown in Fig. 3.1. Now we aggregate the neurons allocated within each layer l, $l = 1, 2, ..., L$, in one place but the information associated with each place $\dddot{x}_{i(l-1)}$ will be delivered to the transition Z_l by tokens. Each token is associated with one neuron.

The new input place will be denoted $\dddot{X}_{(l-1)}$ and the output places will be denoted $\dddot{X}_{(l)}$. Now let us introduce an extra input-output place $\dddot{X}'_{(l)}$ to each transition Z_l - this place will be responsible for control of the transition action. The new structure of the generalized net is shown in Fig. 3.4.

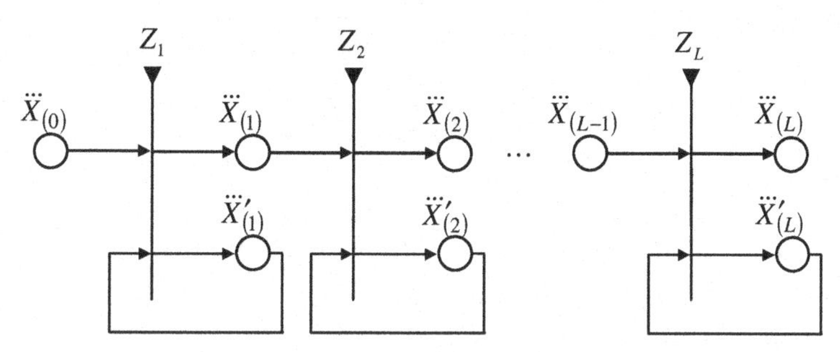

Fig. 3.4 The generalized net representation of the feedforward neural network with layer aggregation

The new structure of the generalized net of the aggregated neural network consists of L transitions, and each transition has the following form

$$Z''_l = \left\langle \left\{ \dddot{X}_{(l-1)}, \dddot{X}'_{(l)} \right\}, \left\{ \dddot{X}_{(l)}, \dddot{X}'_{(l)} \right\}, \tau_l, \tau'_l, r''_l, M''_l, \square''_l \right\rangle \tag{3.70}$$

where $l = 1, 2, ..., L$,

$\left\{ \dddot{X}_{(l-1)}, \dddot{X}'_{(l)} \right\}$
is the set of input places of the l-th transition,

$\left\{ \dddot{X}_{(l)}, \dddot{X}'_{(l)} \right\}$
is the set of output places of the l -the transition,

τ_l
is the time when the l -th transition is fired out,

while $\tau_1 = T$, $\tau_l = T + \sum_{k=2}^{l} \tau'_{k-1}$,

τ'_l
is duration of activity of the l -th transition,

r_l''

is an index matrix, which denotes the l-th transition condition of tokens from the transition's inputs $\{\dddot{X}_{(l-1)}, \dddot{X}'_{(l)}\}$ to its outputs $\{\dddot{X}_{(l)}, \dddot{X}'_{(l)}\}$, and has the following form:

$$r_l'' = \begin{array}{c|cc} & \dddot{X}_{(l)} & \dddot{X}'_{(l)} \\ \hline \dddot{X}_{(l-1)} & false & true \\ \dddot{X}'_{(l)} & V_l & \neg V_l \end{array} \tag{3.71}$$

where

$$V_l = \begin{cases} true & \text{if not all tokens } \alpha_{i(l-1)},\ i(l-1) = 1, 2, \ldots, N(l-1), \\ & \quad \text{entered the transition } Z_l \\ false & \text{otherwise} \end{cases} \tag{3.72}$$

M_l''

indicates an index matrix describing the capacities of transition's arcs:

$$M_l'' = \begin{array}{c|cc} & \dddot{X}_{(l)} & \dddot{X}'_{(l)} \\ \hline \dddot{X}_{(l-1)} & 0 & 1 \\ \dddot{X}'_{(l)} & 1 & 1 \end{array} \tag{3.73}$$

$\square_l'' = \vee\left(\dddot{X}_{(l-1)}, \dddot{X}'_{(l-1)}\right)$

means that at least one place must contain one token.

The new generalized net has the following form:

$$GN3 = \left\langle \langle A, *, \pi_X, c, g, \Theta_1, \Theta_2 \rangle, \langle K, *, \Theta_K \rangle, \langle T, *, t^* \rangle, \langle Y, \Phi, b \rangle \right\rangle \tag{3.74}$$

where

$A = \{Z_1, Z_2, \ldots, Z_L\}$

is a set of transitions,

π_X

is a function describing the priorities of the places in the following way:

$$pr_1\{Z_1, Z_2, \ldots, Z_L\} = \left\{\dddot{X}_{(0)}, \dddot{X}_{(1)}, \dddot{X}'_{(1)}, \dddot{X}_{(2)}, \dddot{X}'_{(2)} \ldots, \dddot{X}_{(L-1)}, \dddot{X}'_{(L)}\right\} \tag{3.75}$$

$$pr_2\{Z_1, Z_2, \ldots, Z_L\} = \left\{\dddot{X}_{(1)}, \dddot{X}'_{(1)}, \dddot{X}_{(2)}, \dddot{X}'_{(2)} \ldots, \dddot{X}_{(L)}, \dddot{X}'_{(L)}\right\} \tag{3.76}$$

c_l
is a function describing the capacities of the places

$$c_l = \begin{cases} N(l) & \text{for } \dddot{X}_{(l)} \\ N(l) & \text{for } \dddot{X}'_{(l)} \end{cases}, \quad l = 0, 1, 2, \ldots, L \tag{3.77}$$

g
is a function that calculates the truth values of the predicates of the transition conditions, and is related to (3.72),

Θ_1
is a function giving the next time-moment when the transitions can be again activated; for a single simulation this time is not valid, while for performing the simulation for a set of inputs the function has the form $\Theta_1(t_l) = t_l'$, $l = 1, 2, \ldots, L$, (with $pr_3 Z_0 = T$), where

$$t_l = T + (p-1)\sum_{k=1}^{L} pr_4 Z_k + pr_3 Z_{l-1} = T + (p-1)\sum_{k=1}^{L} \tau_k' + pr_3 Z_{l-1} \tag{3.78}$$

and

$$t_l' = t_l + \sum_{k=1}^{L} pr_4 Z_k = t_l + \sum_{k=1}^{L} \tau_k' \tag{3.79}$$

$$t' \in [T, T + t^*] \text{ and } t \le t' \text{ for } p = 1, 2, \ldots, P,$$

Θ_2
is a function yielding the duration of activity of each transition Z_l

$$\Theta_2(t_l) = t_l'', \; l = 1, 2, \ldots, L \tag{3.80}$$

where t_l is described by (3.78) and

$$t_l'' = pr_4 Z_l = \tau_l' \tag{3.81}$$

the value of this function is calculated at the moment when the transition starts functioning,

K
is the set of tokens entering the generalized net, here $N(0)$ tokens enter the net,

$$K = \{\alpha_{1(0)}, \alpha_{2(0)}, \ldots, \alpha_{N(0)}\}, \tag{3.82}$$

$\Theta_K = T$
is the time when the first token $\alpha_{1(0)}$ enters the net, and it is the time when the generalized net starts functioning,

t^*
describes the period, during which the generalized net is functioning; for a single simulation $t^* = \sum_{l=1}^{L} \tau'_l$, while for a set of inputs this time is described by $t^* = P \sum_{l=1}^{L} \tau'_l$, where P denotes the number of samples,

$$Y = \{ y(\alpha_{1(0)}), y(\alpha_{2(0)}), ..., y(\alpha_{N(0)}) \} \tag{3.83}$$

is the set of all initial characteristics of the tokens that enter the place $\dddot{X}_{(0)}$, and the entering tokens have the following characteristics

$$y(\alpha_{i(0)}) = \langle NN1, N(0), N(1), i(0), x_{i(0)} \rangle \tag{3.84}$$

for $i(0) = 1,2,...,N(0)$, where $NN1$, $N(0)$, $N(1)$ - are described as before, and $x_{i(0)}$ - is the value of the i -th input to the network,

Φ
the characteristic function generating the new characteristic of the new token can have a very complex form; the simplification of the characteristic (3.84) causes that the lacking elements $imout_{i(0)}$, $F_{(1)}$ and $imout_{i(0)}$ must be the integral parts of Φ ; let the function Φ be described in the following algorithmic form:

1) let $l = 0$
2) $l \leftarrow l + 1$
3) $i(l) = 0$
4) $i(l) \leftarrow i(l) + 1$
5) construct the index matrix

$$imX_{i(l-1)} = [0, ..., 0, x_{i(l-1)}, 0, ..., 0]^T \tag{3.85}$$

6) the token $\alpha_{i(l-1)}$, associated with the place $\dddot{X}_{(l)}$ of the characteristic

$$y(\alpha_{i(l-1)}) = \langle NN1, N(l-1), i(l-1), x_{i(l-1)} \rangle \tag{3.86}$$

(where $NN1$, $N(l-1)$ are described as before, and $x_{i(l-1)}$ is the value of the i -th input to the l -th transition), enters the place $\dddot{X}_{(l)}$, and according to the transition condition (3.71) goes to the place $\dddot{X}'_{(l)}$ (Step 7) if $i(l-1) \le N(l-1)$ otherwise goes to the place $\dddot{X}_{(l)}$ (Step 10)

7) construct the following elements:

$$imX_{(l-1)} \leftarrow imX_{(l-1)} + imX_{i(l-1)} \tag{3.87}$$

where $imX_{i(l-1)} = [0,...,0, x_{i(l-1)}, 0,...,0]^T$

$$imW_{(l-1)} \leftarrow imW_{(l-1)} + imW_{i(l-1)} \tag{3.88}$$

where

$$imW_{i(l-1)} = \begin{array}{c|cccc} & \dddot{x}_{1(l)} & \dddot{x}_{2(l)} & \cdots & \dddot{x}_{N(l)} \\ \hline \dddot{x}_{i(l-1)} & w_{i(l-1)1(l)} & w_{i(l-1)2(l)} & \cdots & w_{i(l-1)N(l)} \end{array} \tag{3.89}$$

$$imout_{(l-1)} \leftarrow imout_{(l-1)} + imout_{i(l-1)}$$

where

$$imout_{i(l-1)} = \begin{array}{c|cccc} & \dddot{x}_{1(l)} & \dddot{x}_{2(l)} & \cdots & \dddot{x}_{N(l)} \\ \hline \dddot{x}_{i(l-1)} & x_{i(l-1)}\, w_{i(l-1)1(l)} & x_{i(l-1)}\, w_{i(l-1)2(l)} & \cdots & x_{i(l-1)}\, w_{i(l-1)N(l)} \end{array}$$

8) the elements (3.87) and (3.88) allow for describing the characteristic of the token $\alpha'_{i(l)}$ associated with the place $\dddot{X}'_{(l)}$ as follows

$$y(\alpha'_{i(l)}) = \langle NN1, N(l-1), N(l), imX_{(l-1)}, imW_{(l-1)} \rangle \tag{3.90}$$

9) go to Step 4

10) for $j = 1,2,...,N(l)$ compute

$$imX_{j(l)} = f_{j(l)}\left(\sum_{i(l-1)=1}^{N(l-1)} imout_{i(l-1)} \right) \tag{3.91}$$

where

$$F_{(l)} = \left[f_{1(l)}\left(\sum_{i(l-1)=1}^{N(l-1)} x_{i(l-1)}\, w_{i(l-1)1(l)} \right), f_{2(l)}\left(\sum_{i(l-1)=1}^{N(l-1)} x_{i(l-1)}\, w_{i(l-1)2(l)} \right),..., \right.$$
$$\left. ..., f_{N(l)}\left(\sum_{i(l-1)=1}^{N(l-1)} x_{i(l-1)}\, w_{i(l-1)N(l)} \right) \right]^T \tag{3.92}$$

for $j(l)=1,2,...,N(l)$ generate the new tokens $\alpha_{j(l)}$ having the following characteristics

$$y(\alpha_{j(l)})=\langle NN1, N(l), j(l), x_{j(l)}\rangle \tag{3.93}$$

where $x_{j(l)}$ is the j-th element of the index matrix

$$imout_{(l-1)} \leftarrow imout_{(l-1)} + imout_{i(l-1)}$$

$$imX_{(l)} = [x_{1(l)}, x_{2(l)}..., x_{N(l)}]^T \tag{3.94}$$

11) if $l<L$ go to Step 2, otherwise go to Step 12

12) the end of generation of the output from the network

$$imX_{(L)} = [x_{1(L)}, x_{2(L)}..., x_{N(L)}]^T \tag{3.95}$$

$b(\alpha_{(l)})=b(\alpha'_{(l)})=1$ for $l=0,1,2,...,L$.

There is a great difference between these two executions. Execution I is characterised by introduction of only one token but with much extended characteristic, due to this fact the information entering each transition is delivered in parallel. For the execution II we have introduced the extra places with extra tokens. In this way the information of each token enters each transition in a sequential way.

3.4 Case C – Full Aggregation

In this case we do not distinguish any subsystems within the network. The network is represented by the following form

$$X(L)=G(WW(L), X(0)) \tag{3.96}$$

where

$$WW(L)=[W_{(1)}, W_{(2)},..., W_{(L)}]^T \tag{3.97}$$

$$W(l)=[w_{1(l)}, w_{2(l)},..., w_{N(l)}]^T \tag{3.98}$$

and $w_{i(l-1)} = [w_{i(l-1)1(l)}, w_{i(l-1)2(l)},..., w_{i(l-1)N(l)}]^T$

while $w_{i(l-1)j(l)}$ are the weights connecting the $i(l-1)$-st neuron from the $(l-1)$-st layer with the $j(l)$-th from the l-th layer, $l=1,2,...,L$, $j(l)=1,2,...,N(l)$.

In this section we aggregate all transitions Z_l, $l=1,2,...,L$, in only one transition Z, and all places are aggregated in only three places $\ddot{X}_1, \ddot{X}_2, \ddot{X}_3$ as shown in

Fig. 3.5. Similarly as in execution II in Sect. 3.3 the tokens represent the neurons of the neural network.

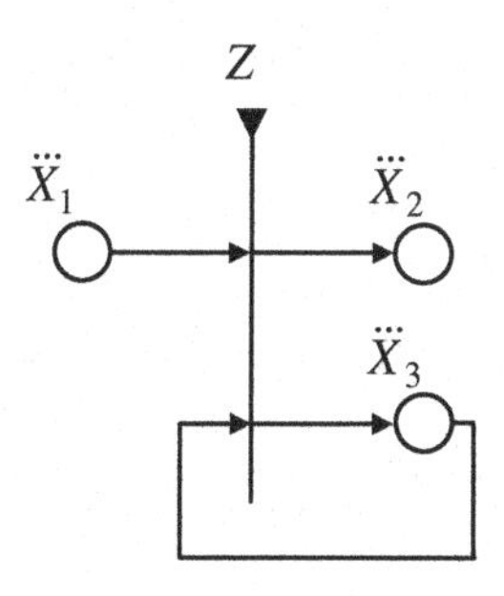

Fig. 3.5 The generalized net representation of the aggregated feedforward neural network

The generalized net representation has the following formal description

$$GN4 = \left\langle \left\langle \{Z\}, *, \pi_X, c, g, \Theta_1, \Theta_2 \right\rangle, \left\langle K, *, \Theta_K \right\rangle, \left\langle T, *, t* \right\rangle, \left\langle Y, \Phi, b \right\rangle \right\rangle \tag{3.99}$$

where the transition has the form

$$Z = \left\langle \left\{\dddot{X}_1, \dddot{X}_3\right\}, \left\{\dddot{X}_2, \dddot{X}_3\right\}, \tau, \tau', r, M, \square \right\rangle \tag{3.100}$$

where

$\left\{\dddot{X}_1, \dddot{X}_3\right\}$

is the set of input places of the transition,

$\left\{\dddot{X}_2, \dddot{X}_3\right\}$

is the set of output places of the transition,

τ

is a time when the transition is fired out, while $\tau = T$,

τ'

is duration of activity of the transition,

r

is an index matrix, which denotes the transition condition of tokens from the transition's inputs $\left\{\dddot{X}_1, \dddot{X}_3\right\}$ to its outputs $\left\{\dddot{X}_2, \dddot{X}_3\right\}$, and has the following form:

$$r = \begin{array}{c|cc} & \dddot{X}_2 & \dddot{X}_3 \\ \hline \dddot{X}_1 & \mathit{false} & \mathit{true} \\ \dddot{X}_3 & V & \neg V \end{array} \tag{3.101}$$

where

$$V = \begin{cases} true & \text{if not all tokens } \alpha_{i(l-1)},\ i(l-1)=1,2,...,N(l-1), \\ & \quad l=1,2,...,L,\ \text{entered the transition } Z \\ false & \text{otherwise} \end{cases}, \tag{3.102}$$

M
indicates an index matrix describing the capacities of transition's arcs:

$$M = \begin{array}{c|cc} & \dddot{X}_2 & \dddot{X}_3 \\ \hline \dddot{X}_1 & 0 & 1 \\ \dddot{X}_3 & 1 & 1 \end{array} \tag{3.103}$$

$\Box = \vee(\dddot{X}_1, \dddot{X}_3)$
means that at least one place must contain one token,

π_X
describes the priorities of the places in the following way:

$$pr_1\{Z\} = \{\dddot{X}_1, \dddot{X}_3\} \tag{3.104}$$

$$pr_2\{Z\} = \{\dddot{X}_2, \dddot{X}_3\} \tag{3.105}$$

c
describes the capacities of the places as follows

$$c(\dddot{X}_1) = c(\dddot{X}_3) = \sum_{l=0}^{L} N(l)$$
$$c(\dddot{X}_2) = N(L) \tag{3.106}$$

g
is a function which calculates the truth values of the predicates of the transition conditions and is related to calculation of (3.102),

Θ_1
is a function giving the next time-moment when the transitions can be again activated; for a single simulation this time is not valid, while for performing the simulation for a set of inputs the function has a form $\Theta_1(t) = t + \tau'$, where $t = T + (p-1)\tau'$, for $p = 1,2,...,P$,

$\Theta_2 = \tau'$

is a function giving the duration of activity of the transition Z,

K

is the set of tokens entering the generalized net, in this case $\sum_{l=0}^{L} N(l)$ tokens enter the net,

$$K = \{\alpha_{1(0)}, \alpha_{2(0)}, \ldots, \alpha_{N(0)}\} \tag{3.107}$$

$\Theta_K = T$

is a time when the first token $\alpha_{1(0)}$ enters the net,

t^*

describes the period within the generalized net is functioning; for a single simulation $t^* = \tau'$, while for a set of inputs this time is described by $t^* = P\tau'$, where P denotes the number of samples,

$$Y = \{y(\alpha_{1(0)}), y(\alpha_{2(0)}), \ldots, y(\alpha_{N(0)})\} \tag{3.108}$$

denotes the set of all initial characteristics of the tokens which enter the place $\dddot{X}_1$, the entering tokens have the following characteristic

$$y(\alpha_{i(0)}) = \langle NN1,\ N(0),\ i(0),\ x_{i(0)} \rangle \tag{3.109}$$

for $i(0) = 1, 2, \ldots, N(0)$,

where

$NN1$, $N(0)$

are described as before,

$x_{i(0)}$

is the value of the i-th input to the network,

Φ

the characteristic function generating the new characteristics of the new tokens in the considered case has a very complex form; the whole processing in this case of generalized net is performed during the tokens $\alpha'_{i(l)}$ associated with the place $\dddot{X}_3$ enter the transition, and the new characteristics are generated; the function Φ will be described in the following algorithmic form:

1) $i(0) = 0$

2) $i(0) \leftarrow i(0) + 1$

3) construct the index matrix

$$imX_{i(0)} = [0,...,0, x_{i(0)}, 0,...,0]^T \tag{3.110}$$

4) if $i(0) > N(0)$ go to Step 8, otherwise the token $\alpha_{i(0)}$, associated with the place $\dddot{X}_1$, having the following characteristic

$$y(\alpha_{i(0)}) = \langle NN1, N(0), i(0), x_{i(0)} \rangle \tag{3.111}$$

due to the transition condition (3.101) passes the transition and next enters the place $\dddot{X}_3$,

5) construct the following elements:

$$imX_{(0)} \leftarrow imX_{(0)} + imX_{i(0)} \tag{3.112}$$

where $imX_{i(0)} = [0,...,0, x_{i(0)}, 0,...,0]^T$

$$imW_{(0)} \leftarrow imW_{(0)} + imW_{i(0)} \tag{3.113}$$

where

$$imW_{i(0)} = \begin{array}{c|cccc} & \dddot{x}_{1(1)} & \dddot{x}_{2(1)} & \cdots & \dddot{x}_{N(1)} \\ \hline \dddot{x}_{i(0)} & w_{i(0)1(1)} & w_{i(0)2(1)} & \cdots & w_{i(0)N(1)} \end{array}$$

$$imout_{(0)} \leftarrow imout_{(0)} + imout_{i(0)} \tag{3.114}$$

where

$$imout_{i(0)} = \begin{array}{c|cccc} & \dddot{x}_{1(1)} & \dddot{x}_{2(1)} & \cdots & \dddot{x}_{N(1)} \\ \hline \dddot{x}_{i(0)} & x_{i(0)}\, w_{i(0)1(1)} & x_{i(0)}\, w_{i(0)2(1)} & \cdots & x_{i(0)}\, w_{i(0)N(1)} \end{array}$$

6) the elements (3.112), (3.113)) and (3.114) allow for describing the characteristic of the token $\alpha'_{i(0)}$ associated with the place $\dddot{X}_3$ as follows

$$y(\alpha'_{i(0)}) = \langle NN1,\ N(0),\ i(1),\ imX_{(0)},\ imW_{(0)},\ imout_{(0)} \rangle \tag{3.115}$$

7) go to Step 4

8) for $j(1) = 1, 2, ..., N(1)$ compute

$$imX_{j(1)} = f_{j(1)}\left(\sum_{i(0)=1}^{N(0)} imout_{i(0)} \right) \tag{3.116}$$

where

$$F_{(1)} = \left[f_{1(1)}\left(\sum_{i(0)=1}^{N(0)} x_{i(0)} w_{i(0)1(1)}\right), f_{2(1)}\left(\sum_{i(0)=1}^{N(0)} x_{i(0)} w_{i(0)2(1)}\right), \ldots, \right.$$
$$\left. \ldots, f_{N(1)}\left(\sum_{i(0)=1}^{N(0)} x_{i(0)} w_{i(0)N(1)}\right)\right]^T \quad (3.117)$$

9) $l = 1$

10) $l \leftarrow l + 1$

11) $i(l) = 0$

12) $i(l) \leftarrow i(l) + 1$

13) if $l > L$ go to Step 18 otherwise continue as follows:
the token $\alpha'_{i(l-1)}$ starts from place $\dddot{X}_3$, than passes the transition Z (according to condition (3.71)) and goes again to the place $\dddot{X}_3$ (or to the place $\dddot{X}_2$ for $l > L$); the token has the following characteristic

$$y(\alpha'_{i(l-1)}) = \langle NN1,\, N(l-1),\, i(l-1),\, imX_{(l-1)},\, imW_{(l-1)},\, imout_{(l-1)} \rangle \quad (3.118)$$

14) compute the elements of (3.118):

$$imX_{(l-1)} \leftarrow imX_{(l-1)} + imX_{i(l-1)} \quad (3.119)$$

where $imX_{i(l-1)} = [0, \ldots, 0, x_{i(l-1)}, 0, \ldots, 0]^T$

$$imW_{(l-1)} \leftarrow imW_{(l-1)} + imW_{i(l-1)} \quad (3.120)$$

where

$imW_{i(l-1)} =$

	$\dddot{x}_{1(l)}$	$\dddot{x}_{2(l)}$	$\cdots$	$\dddot{x}_{N(l)}$
$\dddot{x}_{i(l-1)}$	$w_{i(l-1)1(l)}$	$w_{i(l-1)2(l)}$	$\cdots$	$w_{i(l-1)N(l)}$

$$imout_{(l-1)} \leftarrow imout_{(l-1)} + imout_{i(l-1)} \quad (3.121)$$

where

$imout_{i(l-1)} =$

	$\dddot{x}_{1(l)}$	$\dddot{x}_{2(l)}$	$\cdots$	$\dddot{x}_{N(l)}$
$\dddot{x}_{i(l-1)}$	$x_{i(l-1)} w_{i(l-1)1(l)}$	$x_{i(l-1)} w_{i(l-1)2(l)}$	$\cdots$	$x_{i(l-1)} w_{i(l-1)N(l)}$

15) if $i < N(l-1)$ go to Step 12, otherwise go to Step 16

16) for $j(l) = 1, 2, \ldots, N(l)$ compute

$$imX_{j(l)} = f_{j(l)}\left(\sum_{i(l-1)=1}^{N(l-1)} imout_{i(l-1)}\right) \tag{3.122}$$

where

$$F_{(l)} = \left[f_{1(l)}\left(\sum_{i(l-1)=1}^{N(l-1)} x_{i(l-1)} w_{i(l-1)1(l)}\right), f_{2(l)}\left(\sum_{i(l-1)=1}^{N(l-1)} x_{i(l-1)} w_{i(l-1)2(l)}\right), \ldots, f_{N(l)}\left(\sum_{i(l-1)=1}^{N(l-1)} x_{i(l-1)} w_{i(l-1)N(l)}\right)\right]^T \tag{3.123}$$

17) go to Step 10

18) the end of generating the output of the network

$$imX_{(L)} = \left[x_{1(L)}, x_{2(L)}, \ldots, x_{N(L)}\right]^T \tag{3.124}$$

$b(\alpha_{(l)}) = b(\alpha'_{(l)}) = 1$ for $l = 0, 1, 2, \ldots, L$.

In the previous sections of this chapter we have described the concept of generalized nets used for representing the functioning of the multilayer neural networks, the simulation process of this class of networks, and in somehow informal way we have applied many of the sophisticated tools of the generalized nets theory.

The generalized nets theory, as described by Atanassov (1991, 1992, 2007) contains many different operations and relations over the transitions, tokens as well as over the characteristics of tokens. Atanassov introduced and described in details six types of operators: global operators, local operators, hierarchical operators, reducing operators, extending operators, dynamic operators.

3.5 System of Neural Networks

In this section we would like to show several properties of some different operators applied to a system of neural networks. The basic properties of these operators are discussed in detail, from the theoretical point of view, in Atanassov (1991, 1992, 2007). The operators have a major theoretical and practical value because they allow for studying the properties and behaviour of the generalized nets models.

The forms of the generalized nets described earlier in this section can be used to describe the simulation process of a system of neural networks. The proper operators allow changing the structure of the generalized net representation of the whole neural networks system, the number of transitions, and the number of the tokens as well as the characteristics of the tokens.

Let us consider the system of neural networks shown in Fig. 3.6. The considered system of neural networks consists of three subsystems.

The first subsystem is built of m neural networks denoted by $N_{1,i}$, $i = 1, 2, ..., m$. The external signals of this subsystem, denoted by

$$\{X_{0,1}, X_{0,2}, ..., X_{0,m}\} \tag{3.125}$$

enter the neural networks, $N_{1,i}$, $i = 1, 2, ..., m$, where the processing is performed, and the following outputs of this subsystem are generated

$$\{X_{1,1}, X_{1,2}, ..., X_{1,m}\} \tag{3.126}$$

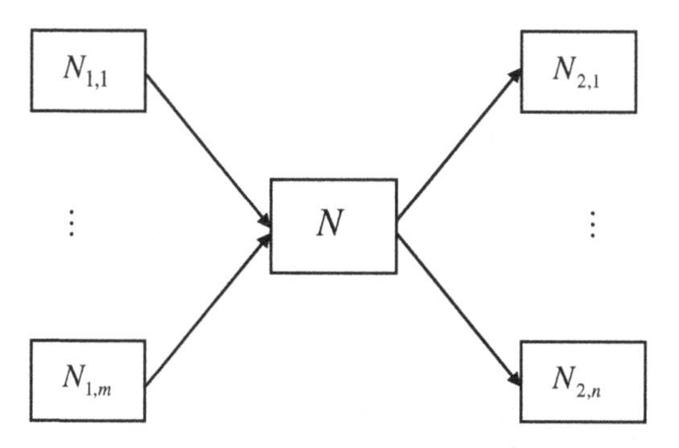

Fig. 3.6 A system of neural networks

The second subsystem is limited to only one neural network denoted by N, the inputs to this subsystem are the outputs of the first subsystem, i.e.

$$\{X_{1,1}, X_{1,2}, ..., X_{1,m}\} \tag{3.127}$$

while the outputs are represented by

$$\{X_{2,1}, X_{2,2}, ..., X_{2,m}\}. \tag{3.128}$$

The third subsystem has a similar form as the first subsystem, namely is built of n neural networks denoted by $N_{2,j}$, $j = 1, 2, ..., n$. The outputs of the second subsystem (128) constitute the inputs to the third subsystem, $N_{2,j}$, $j = 1, 2, ..., n$, and after processing generate the following outputs

$$\{X_{3,1}, X_{3,2}, ..., X_{3,n}\}. \tag{3.129}$$

From the generalized nets theory point of view the first subsystem (3.125) - (3.126) is described by the following sets of elements:
the set of transitions

$$\{Z_{1,1}, Z_{1,2},..., Z_{1,m}\} \tag{3.130}$$

the set of input places

$$\{\dddot{X}_{0,1}, \dddot{X}_{0,2},..., \dddot{X}_{0,m}, \dddot{X}'_{1,1}, \dddot{X}'_{1,2},..., \dddot{X}'_{1,m}\} \tag{3.131}$$

the set of output places

$$\{\dddot{X}_{1,1}, \dddot{X}_{1,2},..., \dddot{X}_{1,m}, \dddot{X}'_{1,1}, \dddot{X}'_{1,2},..., \dddot{X}'_{1,m}\}. \tag{3.132}$$

The generalized net representation of the subsystem (3.127) - (3.128) is described by a single transition Z with a set of input places

$$\{\dddot{X}_1, \dddot{X}'_2\} \tag{3.133}$$

and a set of output places

$$\{\dddot{X}_2, \dddot{X}'_2\}. \tag{3.134}$$

The generalized net representation of the subsystem (3.129) - (3.130) is described by the following sets:
the set of transitions

$$\{Z_{2,1}, Z_{2,2},..., Z_{2,n}\} \tag{3.135}$$

the set of input places

$$\{\dddot{X}_{2,1}, \dddot{X}_{2,2},..., \dddot{X}_{2,n}, \dddot{X}'_{3,1}, \dddot{X}'_{3,2},..., \dddot{X}'_{3,n}\} \tag{3.136}$$

the set of output places

$$\{\dddot{X}_{3,1}, \dddot{X}_{3,2},..., \dddot{X}_{3,n}, \dddot{X}'_{3,1}, \dddot{X}'_{3,2},..., \dddot{X}'_{3,n}\}. \tag{3.137}$$

The set of places $\{\dddot{X}_{3,1}, \dddot{X}_{3,2},..., \dddot{X}_{3,n}\}$ represent the outputs of the whole neural networks system $\{X_{3,1}, X_{3,2},..., X_{3,n}\}$.

The three described subsystems are connected as shown in Fig. 3.6. The generalized net formalism requires the introduction of two additional transitions, namely

$$\{Z_1, Z_2\}. \tag{3.138}$$

The roles of these transitions are rather obvious. The first one is used to transform m places $\dddot{X}_{1,i}$ for $i = 1, 2,..., m$ into one place $\dddot{X}_1$, while the second has to transform one place $\dddot{X}_2$ into n places $\dddot{X}_{2,j}$ for $j = 1, 2,..., n$.

Now let us construct the generalized net representation of the considered system of neural networks. For simplicity, we will consider only the part of the generalized net representation marked by the dashed contour in Fig. 3.7, denoted by G_1.

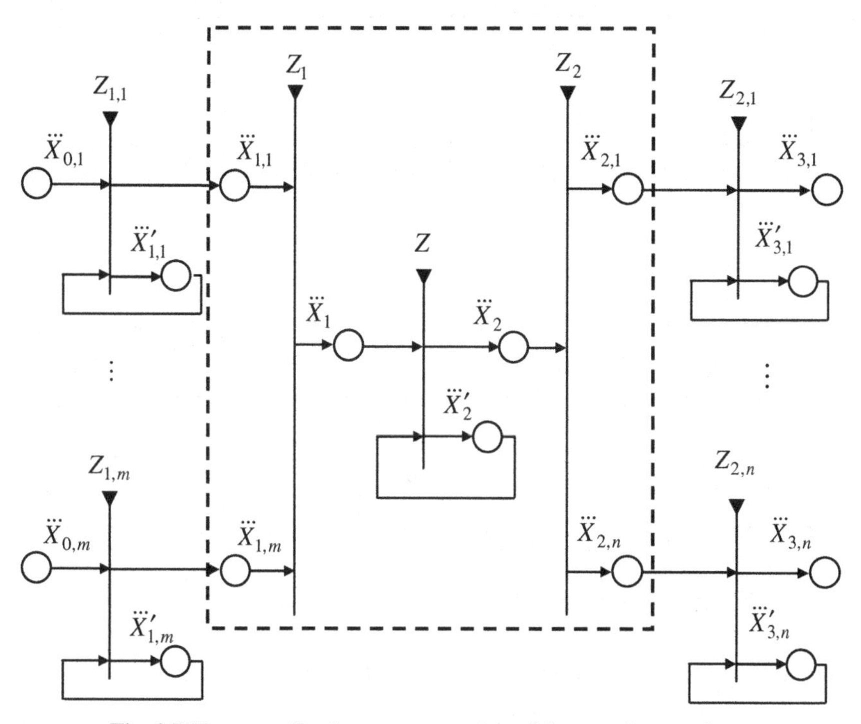

Fig. 3.7 The generalized net representation of the neural network system

In order to illustrate the essential properties of the generalized nets methodology we will transform the net G_1 into a much simpler structure shown in Fig. 3.8, denoted G_2. We can simplify the generalized net G_1 from Fig. 3.7 to the generalized net G_2 from Fig. 3.8 by applying the operators described in details by Atanassov (1991).

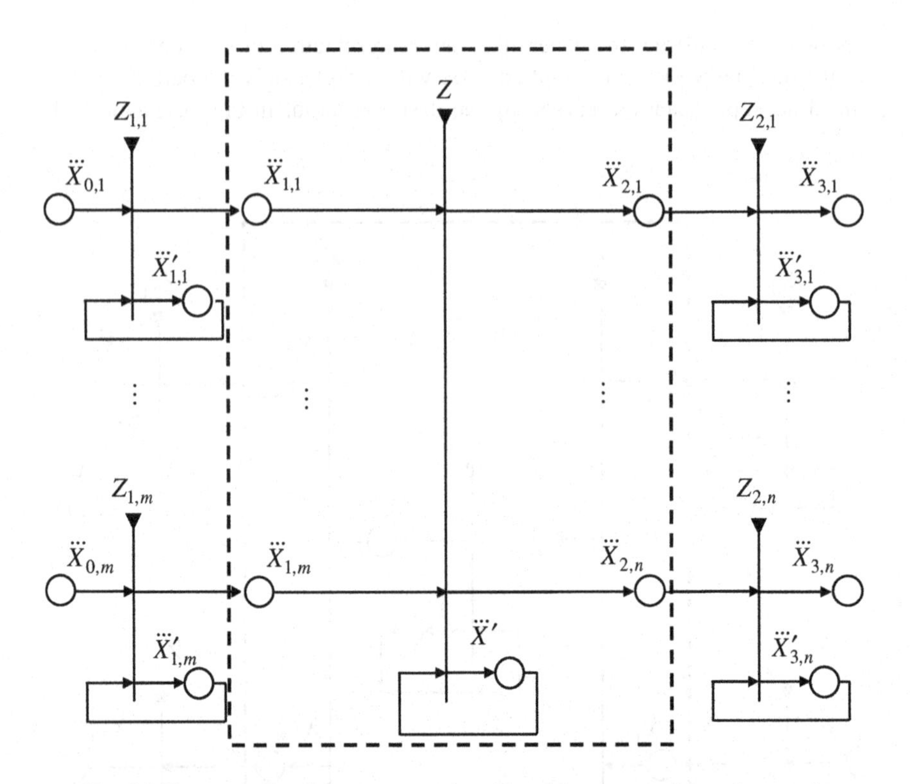

Fig. 3.8 The generalized net representation G_2 of the system of neural networks

In the considered case, the transition Z has the following form

$$Z = \left\langle \left\{ \dddot{X}_{1,1}, \ldots, \dddot{X}_{1,m}, \dddot{X}' \right\}, \left\{ \dddot{X}_{2,1}, \ldots, \dddot{X}_{2,n}, \dddot{X}' \right\}, 0, 1, r, M, \Box \right\rangle \tag{3.139}$$

where the index matrices r and M have the following forms

$$r = \begin{array}{c|cccc} & \dddot{X}_{2,1} & \ldots & \dddot{X}_{2,n} & \dddot{X}' \\ \hline \dddot{X}_{1,1} & false & \ldots & false & true \\ \vdots & \vdots & \ldots & \vdots & \vdots \\ \dddot{X}_{1,m} & false & \ldots & false & true \\ \dddot{X}' & V_1 & \ldots & V_n & \neg V_1 \,\&\ldots\&\, \neg V_n \end{array} \tag{3.140}$$

with

V_i = "the condition of the stopping of the neural network N_i occurred and the process must continue in the i-th direction", $i = 1, 2, \ldots, n$, and

$$M = \begin{array}{c|cccc} & \dddot{X}_{2,1} & \cdots & \dddot{X}_{2,n} & \dddot{X}' \\ \hline \dddot{X}_{1,1} & 0 & \cdots & 0 & 1 \\ \vdots & \vdots & \cdots & \vdots & \vdots \\ \dddot{X}_{1,m} & 0 & \cdots & 0 & 1 \\ \dddot{X}' & 1 & \cdots & 1 & 1 \end{array} \tag{3.141}$$

It is possible to construct several different generalized net representations of the considered case:

- in the first model there are tokens (one or more, but not more than m) that enter input places $\dddot{X}_{1,1}, \dddot{X}_{1,2}, \ldots, \dddot{X}_{1,m}$ of the transition Z, and after that all the tokens are united in one token, whereupon they enter the place $\dddot{X}'$. In the next time step the new token is split into some (one or more, but not more than n) tokens that enter the output places $\dddot{X}_{2,1}, \dddot{X}_{2,2}, \ldots, \dddot{X}_{2,n}$ of the transition Z, in this case the Boolean expression has the form

$$\square = \vee(\dddot{X}_{1,1}, \ldots, \dddot{X}_{1,m}, \dddot{X}' \tag{3.142}$$

- in the second generalized net representation the tokens described above are the same, but now in place $\dddot{X}'$ there is only one fixed token that will stay only in this place; the input tokens entering the transition will be united within one place, and new tokens will be generated and moved to the outputs, while the index matrix transition condition has the following form

$$r = \begin{array}{c|cccc} & \dddot{X}_{2,1} & \cdots & \dddot{X}_{2,n} & \dddot{X}' \\ \hline \dddot{X}_{1,1} & false & \cdots & false & true \\ \vdots & \vdots & \cdots & \vdots & \vdots \\ \dddot{X}_{1,m} & false & \cdots & false & true \\ \dddot{X}' & V_1 & \cdots & V_n & true \end{array} \tag{3.143}$$

while the transition type has the form

$$\square = \wedge\left(\vee(\dddot{X}_{1,1}, \ldots, \dddot{X}_{1,m}, \dddot{X}'), \dddot{X}'\right). \tag{3.144}$$

In both models the capacity of the place $\dddot{X}'$ is equal 1, $c(\dddot{X}') = 1$. It should be noted that in the first case (i.e. without a special token in $\dddot{X}'$) there is other modification of the model, namely it is prohibited to unite the tokens,

- in the third model it is assumed that $c(\dddot{X}')=const \neq 1$ for $c(\dddot{X}')=m$ or $c(\dddot{X}')=m+1$ and the tokens from places $\dddot{X}_{1,1},\ldots,\dddot{X}_{1,m}$ will not be united in one new token in place $\dddot{X}'$, additionally, the tokens will be transferred according to the respective predicates of the following index matrix transition condition

$$r=\begin{array}{c|cccc} & \dddot{X}_{2,1} & \ldots & \dddot{X}_{2,n} & \dddot{X}' \\ \hline \dddot{X}_{1,1} & false & \ldots & false & V_1' \\ \vdots & \vdots & \ldots & \vdots & \vdots \\ \dddot{X}_{1,m} & false & \ldots & false & V_m' \\ \dddot{X}' & V_1 & \ldots & V_n & false \end{array} \tag{3.145}$$

- the fourth model is very similar to the third one, the only difference is that the tokens from places $\dddot{X}_{1,1},\ldots,\dddot{X}_{1,m}$ will not be united in one new token in place $\dddot{X}'$ together with the local token staying in place $\dddot{X}'$; the index matrix transition condition has the following form

$$r=\begin{array}{c|cccc} & \dddot{X}_{2,1} & \ldots & \dddot{X}_{2,n} & \dddot{X}' \\ \hline \dddot{X}_{1,1} & false & \ldots & false & V_1' \\ \vdots & \vdots & \ldots & \vdots & \vdots \\ \dddot{X}_{1,m} & false & \ldots & false & V_m' \\ \dddot{X}' & V_1 & \ldots & V_n & true \end{array} \tag{3.146}$$

  where V_i are the same predicates as above, and V_j' = "the current token (representing a signal from some given neural network) must enter the j-th neural network being is the output of the neural network N", $j=1,2,\ldots,m$.

The representations constructed illustrate the possibilities of application of the generalized net methodology to the simulation process of the system of neural networks. In all the cases considered we used only the simplest forms of the generalized net representations. With these, generalized net representations we can describe the behaviour of a system of neural networks, while these representations do not carry any information of the considered neural network structures.

More complex generalized net descriptions can also represent the neural network structures. In order to apply more advanced generalized nets theory we must specify the detailed forms of transitions. Some hierarchical operators defined over the generalized nets can realize this procedure. We can construct a generalized net with a structure similar to the structure of a given neural network (or system of neural networks). As it was shown in the earlier sections, we can start from a very complex form of the generalized nets, and we may construct much simpler forms of the generalized net representations.

Chapter 4
Learning from Examples

4.1 Introduction

The process of learning from examples (or supervised learning) of multilayer neural networks can be considered when a set $\{input_p, output_p\}$, $p = 1,2,...,P$, is available. The aim is to configure a neural network in such a way as to generate the definite output if the respective input feeds the network. The generation should be done with highest possible accuracy, the requirement that can be written as a difference between the desired $output_p$ and the output generated by the network NN_p, summed over all the available examples P, as

$$E = \sum_{p=1}^{P} \phi\left(output_p - NN_p\left(input_p\right)\right)^2 \le \varepsilon \tag{4.1}$$

where E denotes the performance index of learning (or the error of learning), $\phi(\ )$ is a measure of error, and ε is a prescribed positive small value.

In this book we will consider the Euclidean measure of the learning error

$$E = \frac{1}{2}\sum_{p=1}^{P}\left(output_p - NN_p\left(input_p\right)\right)^2 = \sum_{p=1}^{P} E_p. \tag{4.2}$$

This form of the error is the most widely used one. The rationale came from the theory of signal processing, where energy of the error between two signals has a similar form. The term *energy function* for neural networks was introduced by Hopfield (1982) and was borrowed from statistical mechanics of magnetic systems. Additionally, the form (4.2) together with introducing the sigmoidal activation functions by Hinton and Sejnowski (1983) made the learning of the neural networks as a nonlinear differentiable optimisation problem.

The learning process of the neural networks is considered as an iterative procedure. There are two main ways of updating the weights. The first is based on *the incremental learning*, i.e. the weights are updated after presentation of every pair $\{input_p, output_p\}$, $p = 1,2,...,P$, namely

M. Krawczak: *Multilayer Neural Networks*, SCI 478, pp. 71–93.
DOI: 10.1007/978-3-319-00248-4_4

$$w_p^{new} = w_p^{old} + \Delta w_p \tag{4.3}$$

in such a way that

$$E_p\left(w_p^{old}\right) < E_p\left(w_p^{new}\right). \tag{4.4}$$

The second way is called *batch learning*, and the weights are updated after presentation the whole set of the training pairs, namely

$$w^{new} = w^{old} + \Delta w \tag{4.5}$$

in order to obtain

$$E\left(w^{old}\right) < E_p\left(w^{new}\right). \tag{4.6}$$

In the subsequent material, the *old* values of the weights will be called as *nominal* values and denoted by $\overline{w}$, while the updated values will be denoted by w'.

Additionally, let us remind of the following aspects related to the considered neural networks:

- the network consists of L layers, each layer is labelled by $l = 0,1,2,...,L$, the layer $l = 0$ denotes the external inputs to the network,
- each layer is composed of $N(l)$, $l = 0,1,2,...,L$, neurons, where $N(0)$ denotes the number of inputs,
- the neurons belonging to the $(l-1)$-st layer are connected with the neurons of the l-th layer;
- the activation function of each neuron is defined by one of the following sigmoidal functions

$$f\left(net_{j(l)}\right) = \frac{1}{1+\exp\left(-net_{j(l)}\right)} \tag{4.7}$$

$$f\left(net_{j(l)}\right) = \frac{2}{1+\exp\left(-net_{j(l)}\right)} - 1 \tag{4.8}$$

where

$$net_{j(l)} = net_j(l) = \sum_{i(l-1)=0}^{N(l-1)} w_{i(l-1)j(l)}\, x_{i(l-1)} \tag{4.9}$$

- the learning of the network is understood as an adjustment of the following weights

$$w_{i(l-1)j(l)},\; i(l-1) = 1,2,...,N(l-1),\; j(l) = 1,2,...,N(l),\; l = 1,2,...,L$$

meant to minimize the performance index describing the error of learning

$$E = \frac{1}{2}\sum_{p=1}^{P}\left(D_p - X_p(L)\right)^2 = \frac{1}{2}\sum_{p=1}^{P}\sum_{i(L)=1}^{N(L)}\left(d_{pi(L)} - x_{pi(L)}\right)^2 \tag{4.10}$$

where $D_p = \left[d_{p1}, d_{p2}, \ldots, d_{pN(L)}\right]^T$ is the desired vector of outputs, and

$$X_p(L) = \left[x_{p1(L)}, x_{p2(L)}, \ldots, x_{pN(L)}\right]^T$$

is the response of the network, P denoting the number of the learning examples.

4.2 Delta Learning Rule

The classical approach to the neural networks learning, called the *delta rule*, which gave the foundation for the currently used learning algorithms, was proposed by Widrow and Hoff in 1960. They introduced the *adaptive linear element* (the adaline), an electric device with adjustable resistors as changeable weights, a summing circuit, and a relay, as the *discrete activation function*, realizing the outputs equal either -1 or $+1$ - subject to the summation circuit output. The device is shown in Fig. 4.1.

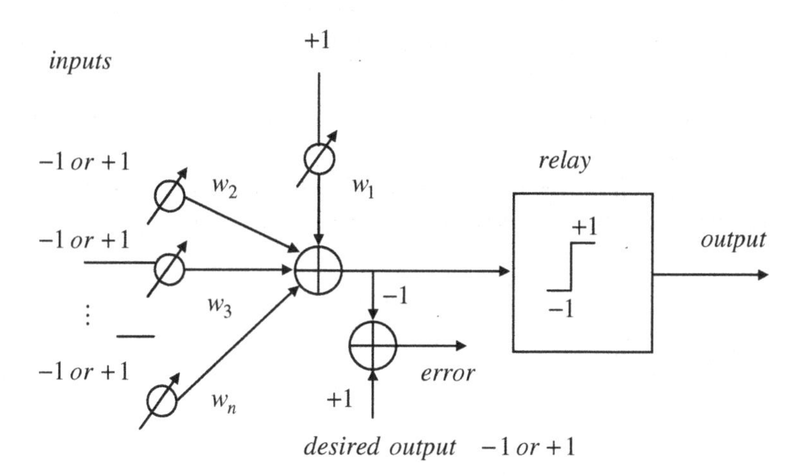

Fig. 4.1 The ADALINE

The signals used in the adaline are all equal -1 or $+1$. It was required to generate a desired $output_p$, when the $input_p = \left[1, x_{p2}, \ldots, x_{pn}\right]^T$ is applied as the input. The problem of adjusting the weights $w_i = 0,1,2,\ldots,n$, was solved by introducing the error between the desired output and the result of the summing point, as

$$error_p = output_p - net_p = output_p - w_p^T x_p \tag{4.11}$$

they proposed the following additional performance index

$$\tilde{E} = \sum_{p=1}^{P} \tilde{E}_p = \frac{1}{2}\sum_{p=1}^{P}\left(output_p - net_p\right)^2 = \frac{1}{2}\sum_{p=1}^{P}\left(output_p - \sum_{i=1}^{n} w_{pi}\, x_{pi}\right)^2 . \tag{4.12}$$

Next, they considered the gradient of $\tilde{E}_p$ with respect to w_{pi}

$$\frac{\partial \tilde{E}_p}{\partial w_{pi}} = \frac{\partial \tilde{E}_p}{\partial net_p}\frac{\partial net_p}{\partial w_{pi}} = -\left(output_p - \sum_{i=1}^{n} w_{pi}\, x_{pi}\right) x_{pi} \tag{4.13}$$

for $i = 1, 2, \ldots, n$,

and for the single weight adjustment they obtained the expression for the weight correction

$$\Delta w_{pi} = -\eta \frac{\partial \tilde{E}_p}{\partial w_{pi}} = \eta\left(output_p - \sum_{i=1}^{n} w_{pi}\, x_{pi}\right) x_{pi} = \eta\, \delta_p\, x_{pi} \tag{4.14}$$

for $i = 1, 2, \ldots, n$,

where the error was denoted by

$$\delta_p = -\frac{\partial \tilde{E}_p}{\partial net_p} = \left(output_p - \sum_{i=1}^{n} w_{pi}\, x_{pi}\right) \tag{4.15}$$

and η is a small positive value. The approach was named the *Widrow and Hoff rule* or the *delta rule.*

Widrow and Hoff proved that the considered error tends to the proper error of learning when the number of iterations goes to infinity, namely

$$\tilde{E} \xrightarrow[\text{number of iterations} \to \infty]{} E = \frac{1}{2}\sum_{p=1}^{P}\left(output_p - \operatorname{sgn}\left(\sum_{i=1}^{n} w_i\, x_i\right)\right)^2 \tag{4.16}$$

where

$$\operatorname{sgn}\left(\sum_{i=0}^{n} w_i\, x_i\right) = \begin{cases} +1 & \text{if } \sum_{i=0}^{n} w_i\, x_i \geq 0 \\ -1 & \text{if } \sum_{i=0}^{n} w_i\, x_i < 0 \end{cases} . \tag{4.17}$$

The use of the sigmoidal function instead of signum function as an activation function in the neuron model, started by Hinton and Sejnowski (1983), allowed for extension of the delta rule for the differentiable activation functions.

Let us consider a single neuron with *the continuous activation function*, shown in Fig. 4.2.

The performance index of learning for all $\{input, output\}$ pairs is expressed by

$$E = \sum_{p=1}^{P} E_p = \frac{1}{2}\sum_{p=1}^{P}\left(output_p - y_p\right)^2 \tag{4.18}$$

where

$$y_p = f\left(net_p\right) = f_p\left(\sum_{i=1}^{n} w_i\, x_{pi}\right)$$

corresponds to the input $x_p = \left[1, x_{p2}, \ldots, x_{pn}\right]^T$

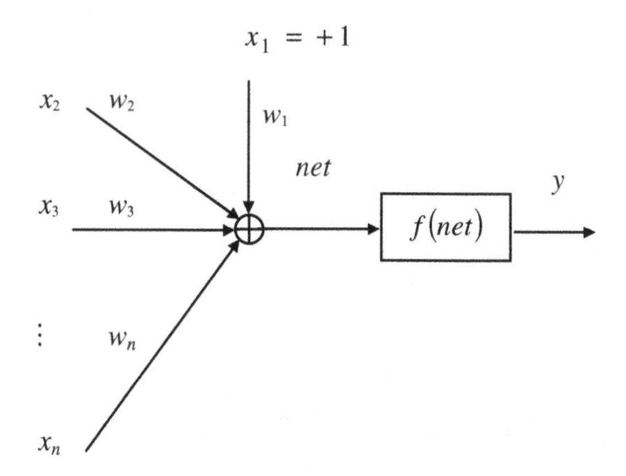

Fig. 4.2 A single neuron with sigmoidal activation function

The function (4.18) can be approximated by expanding up to the second term using the Taylor formula around the point $\overline{w} + \Delta w$ in the following way

$$E\left(\overline{w} + \Delta w\right) = E\left(\overline{w}\right) + \left(\frac{\partial E}{\partial w}\right)^T \Delta w^T + \frac{1}{2}\Delta w^T \frac{\partial^2 E}{\partial w^2}\Delta w \tag{4.19}$$

where Δw is a column vector with elements sufficiently small to ensure the validity of the expansion. For $\overline{w}$ being the minimum point of (4.14) the increment Δw has the following form

$$\Delta w = -\frac{1}{2}\left(\frac{\partial^2 E}{\partial w^2}\right)^{-1}\frac{\partial E}{\partial w}. \tag{4.20}$$

In order to avoid the troublesome computation of the inverse of the Hessian one can replace $\left(\frac{\partial^2 E}{\partial w^2}\right)^{-1}$ by a constant positive value, called the learning rate η, and the weight increment takes the form

$$\Delta w = w' - \overline{w} = -\eta \frac{\partial E}{\partial w} \tag{4.21}$$

For the incremental learning, each weight is updated according to the following formula

$$\begin{aligned}\Delta w_{pi} &= w'_{pi} - \overline{w}_{pi} = -\eta \frac{\partial E}{\partial w_{pi}} = -\eta \frac{\partial E_p}{\partial net_p} \frac{\partial net_p}{\partial w_{pi}} \\ &= \eta \left(output_p - f_p\left(\sum_{i=0}^{n} w_{pi}\, x_{pi}\right)\right) \frac{\partial f_p\left(\sum_{i=1}^{n} w_{pi}\, x_{pi}\right)}{\partial net_p} \frac{\partial f_p\left(\sum_{i=1}^{n} w_{pi}\, x_{pi}\right)}{\partial w_{pi}} \\ &= \eta\, \delta_p \frac{\partial f_p\left(\sum_{i=0}^{n} w_{pi}\, x_{pi}\right)}{\partial w_{pi}}\end{aligned} \tag{4.22}$$

where the delta factor is given by

$$\delta_p = -\frac{\partial E_p}{\partial net_p} = \left(output_p - f_p\left(\sum_{i=1}^{n} w_i\, x_{pi}\right)\right) \frac{\partial f_p\left(\sum_{i=1}^{n} w_i\, x_{pi}\right)}{\partial net_p}. \tag{4.23}$$

For the batch learning, the weights are updated in the following way

$$\begin{aligned}\Delta w_i &= w'_i - \overline{w}_i = -\eta \frac{\partial E}{\partial w_i} = -\eta \frac{\partial E}{\partial net} \frac{\partial net}{\partial w_i} \\ &= \eta \sum_{p=1}^{P} \left(output_p - f\left(\sum_{i=1}^{n} w_i\, x_{pi}\right)\right) \frac{\partial f\left(\sum_{i=1}^{n} w_i\, x_{pi}\right)}{\partial net} \frac{\partial \sum_{p=1}^{P} f\left(\sum_{i=1}^{n} w_i\, x_{pi}\right)}{\partial w_i} \\ &= \eta\, \delta \frac{\partial \sum_{p=1}^{P} f\left(\sum_{i=1}^{n} w_i\, x_{pi}\right)}{\partial w_i}\end{aligned} \tag{4.24}$$

where now the delta factor takes the form

$$\delta = -\frac{\partial E}{\partial net} = \sum_{p=1}^{P} \left(\left(output_p - f_p\left(\sum_{i=1}^{n} w_i \, x_{pi} \right) \right) \frac{\partial f_p\left(\sum_{i=1}^{n} w_i \, x_{pi} \right)}{\partial net_p} \right). \tag{4.25}$$

Let us now derive Equ. 4.23 and 4.25 for the sigmoidal activation functions (4.7) and (4.8).

For the unipolar continuous activation function defined by (4.7), the expressions for the delta factor can be obtained as

$$\delta_p = (d_p - y_p)(1 - y_p)\, y_p \tag{4.23a}$$

$$\delta = \sum_{p=1}^{P} \left((d_p - y_p)(1 - y_p)\, y_p \right); \tag{4.25a}$$

for the bipolar continuous activation function defined by (4.8), the respective formulae for the delta can be obtained as

$$\delta_p = \frac{1}{2}(d_p - y_p)(1 - y_p^2) \tag{4.23b}$$

$$\delta = \frac{1}{2}\sum_{p=1}^{P} \left((d_p - y_p)(1 - y_p^2) \right) \tag{4.25b}$$

where $d_p = output_p$, $y_p = f(net_p)$.

The methodology described by Widrow and Hoff in 1960, i.e. the use of the error, and the application the Euclidean measure of the error learning, and then the application of the gradient method for the weight adjustment, was the basis for *the generalized delta rule* described in the following section.

4.3 Generalized Delta Rule

Limited usefulness of the single-layer neural networks was proved by Minsky and Papert (1969). The only way, nowadays, to overcome such restrictions is to use multilayer neural networks with neurons having differentiable activation functions like the sigmoidal ones. Owing to Rumelhart, Hinton and Williams (1986) the methodology known earlier was refreshed and applied to neural networks. The method is known as *the backpropagation algorithm* or *the backpropagation learning rule*. However, the similar algorithms are well known in the optimisation theory, see e.g. Kelley (1960). From a point of view of the neural network field, the algorithm is based on *the generalized delta rule*, which is the delta rule for differentiable nonlinear activation functions extended for the multilayer case, see e.g. Pedrycz (1998) or Osowski (2000).

$$E = \sum_{p=1}^{P} E_p = \frac{1}{2}\sum_{p=1}^{P}\sum_{j(L)=1}^{N(L)}\left(d_{pj(L)} - x_{pj(L)}\right)^2 . \tag{4.26}$$

In the sequel of this section we will consider the incremental learning of the neural networks. Having the nominal weights $\overline{w}_{i(l-1)j(l)}$, $i(l-1)=1,2,\ldots,\ N(l-1)$, $j(l)=1,2,\ldots,N(l)$, $l=1,2,\ldots,L$, we express the modification of the weights connecting the neurons contained in the last L-th layer with the neurons contained in the $(L-1)$-st layer by

$$\begin{aligned}
\Delta w_{pi(L-1)j(L)} &= w'_{pi(L-1)j(L)} - \overline{w}_{pi(L-1)j(L)} \\
&= -\eta \frac{\partial E_p}{\partial w_{pi(L-1)j(L)}} = -\eta \frac{\partial E_p}{\partial net_{pj(L)}} \frac{\partial net_{pj(L)}}{\partial w_{pi(L-1)j(L)}} \\
&= \eta \left(d_p - f_p\left(\sum_{i(L-1)=1}^{N(L-1)} w_{pi(L-1)j(L)}\, x_{pi(L-1)} \right)\right) \\
&\times \frac{\partial f_p\left(\sum_{i(L-1)=1}^{L-1} w_{pi(L-1)j(L)}\, x_{pi(L-1)} \right)}{\partial net_{pj(L)}} \frac{\partial f_p\left(\sum_{i(L-1)=1}^{L-1} w_{pi(L-1)j(L)}\, x_{pi(L-1)} \right)}{\partial w_{pi(L-1)j(L)}} \\
&= \eta\, \delta_{pj(L)} \frac{\partial f_p\left(\sum_{i(L-1)=1}^{L-1} w_{pi(L-1)j(L)}\, x_{pi(L-1)} \right)}{\partial w_{pi(L-1)j(L)}}
\end{aligned} \tag{4.27}$$

where

$$net_{pj(L)} = \sum_{i(L-1)=1}^{N(L-1)} w_{pi(L-1)j(L)}\, x_{pi(L-1)} \tag{4.28}$$

and the delta factor is described by

$$\begin{aligned}
\delta_{pj(L)} &= -\frac{\partial E_p}{\partial net_{pj(L)}} \\
&= \left(d_p - f_p\left(\sum_{i(L-1)=1}^{L-1} w_{i(L-1)j(L)}\, x_{pi(L-1)} \right)\right) \frac{\partial f_p\left(\sum_{i(L-1)=1}^{L-1} w_{i(L-1)j(L)}\, x_{pi(L-1)} \right)}{\partial net_{pj(L)}} .
\end{aligned} \tag{4.29}$$

By Equ. 4.28, we can calculate the derivative of the activation function with respect to the weight $w_{pi(L-1)j(L)}$ in (4.27) as

$$\frac{\partial f_p\left(\sum_{i(L-1)=1}^{L-1} w_{pi(L-1)j(L)}\, x_{pi(L-1)}\right)}{\partial w_{pi(L-1)j(L)}} = x_{pi(L-1)}\,. \tag{4.30}$$

Substituting (4.30) to (4.27) we get the following expression for the weight changes

$$\Delta w_{pi(L-1)j(L)} = w'_{pi(L-1)j(L)} - \overline{w}_{pi(L-1)j(L)} = \eta\, \delta_{pj(L)}\, x_{pi(L-1)} \tag{4.31}$$

where the delta factor can be rewritten in the shortened way as

$$\delta_{pj(L)} = \left(d_p - x_{pi(L-1)}\right) f'_p\left(net_{pj(L)}\right). \tag{4.32}$$

Let us consider the $(L-1)$-st layer. The changes of the weights associated with this layer, $w_{pi(L-2)j(L-1)}$, are expressed by

$$\begin{aligned}
\Delta w_{pi(L-2)j(L-1)} &= w'_{pi(L-2)j(L-1)} - \overline{w}_{pi(L-2)j(L-1)} \\
&= -\eta \frac{\partial E_p}{\partial w_{pi(L-2)j(L-1)}} \\
&= -\eta \frac{\partial E_p}{\partial net_{pj(L-1)}} \frac{\partial net_{pj(L-1)}}{\partial w_{pi(L-2)j(L-1)}} \\
&= -\eta\, \delta_{pj(L-1)} \frac{\partial net_{pj(L-1)}}{\partial w_{pi(L-2)j(L-1)}}
\end{aligned} \tag{4.33}$$

where the third term in (4.33) is equal

$$\frac{\partial net_{pj(L-1)}}{\partial w_{pi(L-2)j(L-1)}} = x_{pi(L-2)}$$

and so the weight correction can be written as

$$\Delta w_{pi(L-2)j(L-1)} = -\eta\, \delta_{pj(L-1)}\, x_{pi(L-2)}\,. \tag{4.34}$$

The delta factor is defined as before, namely

$$\delta_{pj(L-1)} = -\frac{\partial E_p}{\partial net_{pj(L-1)}}\,.$$

Let us compute the expression for the delta factor in the following way

$$\begin{aligned}\delta_{pj(L-1)} &= -\frac{\partial E_p}{\partial net_{pj(L-1)}} \\ &= -\frac{\partial E_p}{\partial x_{pi(L-1)}}\frac{\partial x_{pi(L-1)}}{\partial net_{pj(L-1)}}\end{aligned} \tag{4.35}$$

where the first term can be derived through the reasoning given below

$$\begin{aligned}&\frac{\partial E_p}{\partial x_{pi(L-1)}} = \\ &= \frac{\partial}{\partial x_{pi(L-1)}}\left[\frac{1}{2}\sum_{j(L)=1}^{N(L)}\left(d_{pj(L)} - x_{pj(L)}\right)^2\right] \\ &= \frac{\partial}{\partial x_{pi(L-1)}}\left[\frac{1}{2}\sum_{j(L)=1}^{N(L)}\left(d_{pj(L)} - f_p\left(\sum_{i(L-1)=1}^{L-1} w_{pi(L-1)j(L)}\, x_{pi(L-1)}\right)\right)^2\right] \\ &= \frac{\partial}{\partial x_{pi(L-1)}}\left[\frac{1}{2}\sum_{j(L)=1}^{N(L)}\left(d_{pj(L)} - f_p\left(net_{pj(L)}\right)\right)^2\right] \\ &= -\sum_{j(L)=1}^{N(L)}\left[\left(d_{pj(L)} - x_{pj(L)}\right)\frac{\partial f_p\left(net_{pj(L)}\right)}{\partial x_{pj(L-1)}}\right] \\ &= -\sum_{j(L)=1}^{N(L)}\left[\left(d_{pj(L)} - x_{pj(L)}\right)\frac{\partial f_p\left(net_{pj(L)}\right)}{\partial net_{pj(L)}}\frac{\partial net_{pj(L)}}{\partial x_{pi(L-1)}}\right] \\ &= -\sum_{j(L)=1}^{N(L)}\left[\left(d_{pj(L)} - x_{pj(L)}\right)f_p'\left(net_{pj(L)}\right)\frac{\partial net_{pj(L)}}{\partial x_{pi(L-1)}}\right].\end{aligned} \tag{4.36}$$

This expression can be simplified by using (4.32) as follows

$$\frac{\partial E_p}{\partial x_{pi(L-1)}} = -\sum_{j(L)=1}^{N(L)}\delta_{pj(L)}\, w_{pi(L-1)j(L)}\,. \tag{4.37}$$

Using (4.37) we can now write the expression for the delta factor as

$$\delta_{pj(L-1)} = f_p'\left(net_{pj(L-1)}\right)\sum_{k(L)=1}^{N(L)}\delta_{pk(L)}\, w_{pj(L-1)k(L)} \tag{4.38}$$

for $j(L-1) = 1, 2, \ldots, N(L-1)$.

The weight modification (4.33) in the $(L-1)$-st layer becomes

$$\Delta w_{pi(L-2)j(L-1)} = w'_{pi(L-2)j(L-1)} - \overline{w}_{pi(L-2)j(L-1)}$$
$$= \eta \; f'_p\left(net_{j(L-1)}\right) x_{pi(L-2)} \sum_{k(L)=1}^{N(L)} \delta_{pk(L)} \; w_{pj(L-1)k(L)} \tag{4.39}$$

for $j(L-1) = 1,2,...,N(L-1)$, $i(L-2) = 1,2,...,N(L-2)$.

For any layer labelled by l, $l = 1,2,...,L-2$, the adjustment of the weights is expressed by the following formula

$$\Delta w_{pi(l-1)j(l)} = w'_{pi(l-1)j(l)} - \overline{w}_{pi(l-1)j(l)}$$
$$= \eta \; f'_p\left(net_{j(l)}\right) x_{pi(l-1)} \sum_{k(l+1)=1}^{N(l+1)} \delta_{pk(l+1} \; w_{pj(l)k(l+1)} \tag{4.40}$$

for $j(l) = 1,2,...,N(l)$, $i(l-1) = 1,2,...,N(l-1)$,

where

$$\delta_{pj(l+1)} = f'_p\left(net_{pj(l+1)}\right) \sum_{k(l+2)=1}^{N(l+2)} \delta_{pk(l+2)} \; w_{pj(l+1)k(l+2)} \; . \tag{4.41}$$

Equ. 4.41 expresses *the generalized delta rule*. The meaning of (4.41) is very simple, namely the computed error is propagated in backward manner from the last layer to any layer below and has a strong influence on the weight adjustment of this layer. The generalized delta factor describes the distribution of the output error among the neurons allocated in other layers. A new interpretation of the recursive formula of the generalized delta factor will be demonstrated in Chapter 7.

4.4 Backpropagation Algorithm

The most popular algorithm for the weight adjustment of the multilayer neural networks, the backpropagation algorithm and its modifications, is based on the generalized delta rule. The algorithm consists of two phases characterised by the opposite flow directions of the signals. For the assumed initial values of the weights, within the first phase, the input to the network is propagated through the network from the layer 0 to the L-th layer. Consequently, the output of the network is generated, and it is next compared with the desired output, forming the error of learning (the performance index of learning). During the second phase, the error is propagated in the opposite direction, from the last layer to the input layer, in order to change the weights in such a way as to decrease the value of the error.

The backpropagation algorithm can be described in the following steps:

1. Assume the given training pairs:

$$\{(x_1, d_1), (x_2, d_2), \ldots, (x_P, d_P)\}$$

2. Initialise the nominal weights:

$$\overline{w}_{i(l-1)j(l)}, \quad l = 1, 2, \ldots, L-1$$

3. Set: $E_p = 0$, η to a small positive number

4. Set: $p = 0$

5. Submit a pattern: $p = p + 1$

$$x_{p1}, x_{p2}, \ldots, x_{pN(0)}, \quad d_{p1}, d_{p2}, \ldots, d_{pN(L)}$$

6. Compute the neurons' outputs:

$$x_{pi(l)}, \; l = 1, 2, \ldots, L, \; i(l) = 1, 2, \ldots, N(l)$$

7. Compute the performance index of learning:

$$E_p = E_p + \frac{1}{2} \sum_{j(L)=1}^{N(L)} \left(d_{pj(L)} - x_{pj(L)}\right)^2$$

8. Compute the delta factors:

$$\delta_{pj(l)}, \quad \text{for } l = L, L-1, \ldots, 1$$

9. Compute the weight corrections:

$$\Delta w_{pi(l-1)j(l)} = \eta \, \delta_{pj(l)} \, x_{pi(l-1)}, \quad \text{for } l = L, L-1, \ldots, 2, 1$$

10. If $p = P$ then go to Step 5 else go to Step 11

11. If $E < E_{MAX}$ then go to Step 12 else go to Step 4

12. STOP

The algorithm described above is based on the incremental learning approach, meaning that the weights are modified after each training input x_p, $p = 1, 2, \ldots, P$, is presented as the network input, as well as each desired output d_p, $p = 1, 2, \ldots, P$, is compared to the network output creating the performance index of learning. In a similar way one can derive the backpropagation algorithm for the batch learning approach.

4.5 Computational Properties of Backpropagation

The backpropagation algorithm is the most popular one for the weight adjustment of the multilayer neural networks. The algorithm is based on the gradient methods and therefore it is slow, and so some modifications were proposed.

In the backpropagation algorithm the change of weight is proportional to the gradient of the performance index with respect to the weight. The Taylor expansion is valid for infinitesimal increments, and the inverse of the Hessian $\left(\frac{\partial^2 E}{\partial w^2}\right)^{-1}$, appearing in (4.20), is replaced by the learning rate η. The learning rate is taken as a positive number; large enough to make a long step in the gradient direction, and small enough to avoid oscillations. In order to ensure the long step and to avoid oscillations it is possible to involve the past weight change in the formula describing the actual weight adjustment (4.40)

$$\begin{aligned}\Delta w_{pi(l-1)j(l)}^{new} &= w'_{pi(l-1)j(l)} - \overline{w}_{pi(l-1)j(l)} \\ &= \eta\, f'_p\left(net_{j(l)}\right) x_{pi(l-1)} \sum_{k(l+1)=1}^{N(l+1)} \delta_{pk(l+1} w_{pj(l)k(l+1)} + \alpha\, \Delta w_{pi(l-1)j(l)}^{old}\end{aligned} \tag{4.42}$$

where $0 < \alpha < 1$ is a positive number called *the momentum rate*, $\Delta w_{pi(l-1)j(l)}^{new}$ is the actual change of the $p_{i(l-1)j(l)}$-th weight, while $\Delta w_{pi(l-1)j(l)}^{old}$ indicates the weight change done in the previous iteration. The term $\alpha\, \Delta w_{pi(l-1)j(l)}^{old}$ is called *the momentum*, and its role is to memorize the shape of the performance index, namely if the surface of the performance index is flat, and the previous step was long, then it is allowed to make the actual step also long, and vice versa.

There are other heuristic ways of accelerating the backpropagation method, reported e.g. by Jacobs (1988). Meanwhile, it seems to be natural to use the well known optimisation methods described e.g. by Findeisen, Szymanowski and Wierzbicki (1972) or by Polak (1971), for example *the conjugate gradient method* (Smagt 1993).

Due to the shape of the sigmoidal functions used as the neuron activation functions the state of each neuron can easily achieve saturation, 0 or 1 for the unipolar activation functions, or -1 or 1 for the bipolar activation functions. In result the expressions for the delta factor, (4.23a) and (4.23b), become almost zero and the learning process can be easily *paralysed*. One of the ways to avoid this phenomenon is to normalize the inputs (Azoff 1994). Another way to avoid the paralysis of a network is to use some heuristics, namely Krawczak (1999b) proposed a fuzzy reinforcement of the backpropagation for keeping the activation of the neurons within a prescribed region, e.g. $(-0.95, +0.95)$ for the bipolar activation functions.

The learning process of a multilayer neural network is a nonlinear optimisation problem. The performance index of learning is a multimodal function with many local minima as well as flat areas. One of the ways to improve the conventional backpropagation algorithm is to apply e.g. the conjugate gradient method, which is based on the construction of a set of directions conjugate to each other in such a way that minimization of the performance index in one direction does not harm the minimization in other directions. For quadratic optimisation problems the conjugate gradient method ensures getting the minimum point within the number of steps equal to the dimension of the problem (Polak 1971). Except for the heuristic methods like *the genetic algorithms* (Michalewicz 1996) applied in the search for the global minimum of the learning error, there are also *stochastic* algorithms (Schoen 1991); *simulated annealing* algorithms (Kirkpatrick et al. 1983, Kirkpatrick 1984); algorithms using *the branch and bound Lipschitz optimisation* methods (Tang and Koehler 1994); or the algorithm based on *the principal component analysis* (Baldi and Hornik 1989). Another way to avoid the problem of local minima is to apply one of the global optimisation techniques, e.g. *the dynamic programming* (Krawczak 1999a, 1999b, 2000a, 2001b, 2002b, 2002c, 2004b), described in the next chapter.

Another problem in the neural network learning is related to the sufficient *number* of *the training examples.* This is a question whether the training set is adequately representative for the considered problem. This is important for both the classification problems and function approximation problem, even the performance index reaches a value close to zero. Additionally, dividing the set of training examples into two subsets; one is used for training while second for testing (Masters 1993) must be done carefully.

One of the main tasks of the learning process is the selection of proper architecture of the considered network, where by the architecture we mean the number of layers and the numbers of neurons allocated in each layer. The architecture of the network is closely related to the number of training examples (Ellacott 1994). The architecture of a multilayer neural network should correspond to the properties of the considered problem; this involves the number of training examples, the number of inputs, the number of outputs and the assumed relation between outputs and inputs (Lawrence et al. 1996). There are well known problems of *generalization* and *overfitting* by neural networks (Baum and Haussler 1989). There are even so called *constructive algorithms* and *pruning algorithms* for building the architecture of neural networks, but in solving any practical problem we cannot avoid performing numerous trials in order to find the satisfactory architecture.

4.6 Generalized Net Description of BP Algorithm

In this section, we will construct the generalized net description of the backpropagation algorithm of multilayer feedforward neural networks, the section is mostly base on works by Krawczak and Aladjov (2002) and Krawczak (2003b, 2003e).

We are interested in modelling the learning capabilities of the neural network, and so we will construct the generalized net representation of the neural networks that can describe the changes of the weights between neurons. In the considerations some of the absent elements will be involved in the description of the proper characteristic function Φ, which generates the new tokens characteristics. Additionally, the inputs of the neural network are treated also as tokens.

Three main parts can be distinguished in the description of the neural network learning process. The first part describes the process of simulation or propagation; in the second part the performance index of learning is introduced, while the third part describes the operations that are necessary to change the states of neurons (by changing the connections – i.e. weights).

Let us consider the generalized net representation of the backpropagation algorithm shown in Fig. 4.3. Each neuron or a group of neurons in the neural network (e.g. a layer or all neurons in the considered network) is represented by a token of α-type. The tokens of this type enter the net through the place $\dddot{X}_1$ and have the following initial characteristics

$$y(\alpha_{i(l)}) = \langle NN1,\ l,\ i(l),\ f_{i(l)} \rangle \tag{4.43}$$

for $i(l) = 1, 2, ..., N(l)$, $l = 0, 1, ..., L$, where

$NN1$
the neural network identifier,

$i(l)$
the number of the token (neuron) associated with the l-th layer,

l
the present layer number,

$$f_{1(l)}\left(\sum_{i(l-1)=1}^{N(l-1)} x_{i(l-1)}\, w_{i(l-1)1(l)} \right) \tag{4.44}$$

is an activation function of the i-th neuron associated with the l-th layer of the neural network.

It should be mentioned that the characteristic (4.43) has a different form than that described by e.g. (3.36), because there we were interested in modelling of the neural network simulation process, when there are adjusted weights between neurons and for a given network input and we are interested in the network output - in that case the tokens represented the stages of the neurons (the values of the neurons' outputs). If we are interested in modelling the learning capabilities of the neural network, then we must construct the generalized net representation of the neural network that can describe the changes of the weights between neurons. Here the inputs of the neural network are treated also as tokens, for which the activation function $f_{i(0)}(\cdot) = 1.0$, for $i(0) = 1, 2, ..., N(0)$.

The basic generalized net description of the backpropagation algorithm contains six transitions, see Fig. 4.3, which will be described one by one.

Every token $\alpha_{i(l)}$, $i(0)=1,2,...,N(l)$, $l=0,1,...,L$, is transferred from the place $\dddot{X}_1$ to the place $\dddot{X}_2$ as well as $\dddot{X}_3$ via the transition Z_1. We assume that the tokens are transferred sequentially according to increasing indexes $i(l)=1,2,...,N(l)$ for given $l=0,1,...,L$, in order to be aggregated with other tokens of the same level l into one new token $\alpha_{(l)}$, representing the whole layer l, according the following conditions of transition Z_1

$$Z_1 = \langle \{\dddot{X}_1, \dddot{X}_2\}, \{\dddot{X}_2, \dddot{X}_3\}, \begin{array}{c|cc} & \dddot{X}_2 & \dddot{X}_3 \\ \hline \dddot{X}_1 & V_{1,2} & V_{1,3} \\ \dddot{X}_2 & V_{2,2} & V_{2,3} \end{array}, \vee(\dddot{X}_1, \dddot{X}_2) \rangle \tag{4.45}$$

where

$V_{1,2} = \neg V_{1,3} =$ "if there is only one token $\alpha_{i(l)}$ in the place $\dddot{X}_1$", i.e.

$$\forall \left(\alpha_{j(l)} \in K_{\dddot{X}_1}\right)\left(pr_2 Y_{\alpha_{i(l)}} \neq pr_2 Y_{\alpha_{j(l)}}, j(l) \neq i(l)\right) \tag{4.46}$$

(where $K_{\dddot{X}_1}$ is a set of all tokens entering the net from the place $\dddot{X}_1$, $i(l), j(l)=1,2,...,N(l)$)

$V_{2,2} =$ "if there is more than one token $\alpha_{i(l)}$ and $\alpha_{j(l)}$ associated with the l-th layer", i.e.

$$\exists \left(\alpha_{i(l)} \in K_{\dddot{X}_1} \,\&\, \alpha_{j(l)} \in K_{\dddot{X}_2}\right)\left(pr_2 Y_{\alpha_{i(l)}} = pr_2 Y_{\alpha_{j(l)}}\right) \tag{4.47}$$

$V_{2,3} =$ "if all tokens $\alpha_{i(l)}$, $i(l)=1,2,...,N(l)$, have been combined into one token" i.e.

$$\begin{gathered} \neg\exists \left(\alpha_{i(l)} \in K_{\dddot{X}_1} \,\&\, \alpha_{k(l)} \in K_{\dddot{X}_2}\right)\left(pr_2 Y_{\alpha_{i(l)}} = pr_2 Y_{\alpha_{k(l)}}\right) \& \\ \neg\exists \left(\alpha_{i(l)} \in K_{\dddot{X}_1} \,\&\, \alpha_{j(l)} \in K_{\dddot{X}_1}\right)\left(pr_2 Y_{\alpha_{i(l)}} = pr_2 Y_{\alpha_{j(l)}}, i \neq j\right). \end{gathered} \tag{4.48}$$

As far as we are not interested in network topologies there is no need of considering the separate neurons or even separate layers, in this way all α-type tokens are aggregated into just one. Here, we are interested in changing the characteristics of the neurons, therefore the whole neural network is represented by one transition Z_1 with three places $\dddot{X}_1, \dddot{X}_2, \dddot{X}_3$.

First, we will aggregate the tokens associated with neurons related to one layer. This aggregation is done in the transition Z_1 in the following manner. For each layer, $l=0,1,...,L$, the characteristics of the separated tokes representing neurons

(4.43) are further processed with the matrices of connection weights, the neuron activation functions, the neuron outputs and so on (as it was shown in details in Chap. 3), in order to construct a new token representing the whole layer.

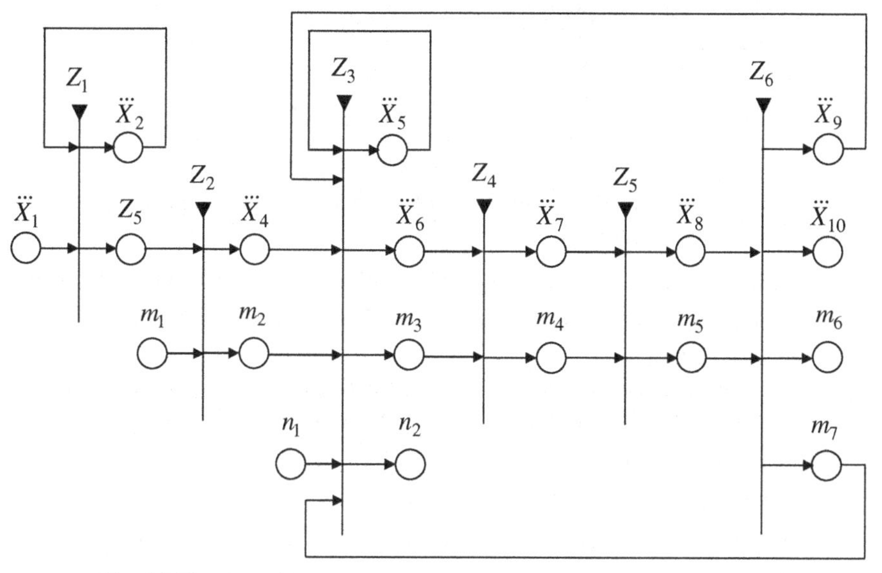

Fig. 4.3 The generalized net rpresentatiom of the backpropagation algorithm

The new token $\alpha_{(l)}$ associated with the l-th layer according to the condition (4.48) is transferred from the place $\dddot{X}_2$ to the place $\dddot{X}_3$, and has the following characteristic

$$y(\alpha_{(l)}) = \langle NN1, l, [1, N(l)], F_{(l)} \rangle \tag{4.49}$$

for $l = 0, 1, 2, \ldots, L$, where

$NN1$
the neural network identifier,

l
the layer number,

$[1, N(l)]$
denotes $N(l)$ tokens (neurons) arranged in a sequence, starting form the first and ending at $N(l)$, associated with the l-th layer,

$$F_{(0)} = [1,1,\ldots,1]^T \tag{4.50}$$

$$F_{(l)} = \left[f_{1(l)}\left(\sum_{i(l-1)=1}^{N(l-1)} x_{i(l-1)} w_{i(l-1)1(l)} \right), f_{2(l)}\left(\sum_{i(l-1)=1}^{N(l-1)} x_{i(l-1)} w_{i(l-1)2(l)} \right), \ldots, f_{N(l)}\left(\sum_{i(l-1)=1}^{N(l-1)} x_{i(l-1)} w_{i(l-1)N(l)} \right) \right]^T$$

is a vector of the activation functions of the neurons associated with the l-th layer of the neural network.

The procedure described above is repeated for all layers, and in result in the place $\dddot{X}_3$ we obtain L tokens, the representation of the procedure of generating the neural network output.

It should be emphasized here that the essential information about neurons' connectivity is contained in the characteristic function Φ of the transition Z_1.

The second transition Z_2 is devoted to introduction of the performance index of the learning process. This kind of information is associated with the β-type token. The token β enters the input place m_1 with the following initial characteristic

$$y(\beta) = \langle NN1, E, E_{\max} \rangle \tag{4.51}$$

where

$NN1$
the neural network identifier,

E
performance index of the neural network learning,

$E_{\max}$
threshold value of the performance index, which must be reached.

The transition Z_2 has the following form

$$Z_2 = \langle \{\dddot{X}_3, m_1\}, \{\dddot{X}_4, m_2\}, \begin{array}{c|cc} & \dddot{X}_4 & m_2 \\ \hline \dddot{X}_3 & true & false \\ m_1 & false & true \end{array}, \wedge(\dddot{X}_3, m_1) \rangle . \tag{4.52}$$

The token $\alpha_{(l)}$, $l = 0,1,2,\ldots,L$, representing the l-th layer obtains the following new characteristic in the place $\dddot{X}_4$

$$y(\alpha_{(l)}) = \langle NN1,\ l,\ [1, N(l)], F_{(l)}, \overline{W}_{(l)} \rangle \tag{4.53}$$

for $l = 0,1,2,...,L$, where

$NN1$
the neural network identifier,

l
the layer number,

$[1, N(l)]$
denotes $N(l)$ tokens (neurons) arranged in a sequence,
starting form the first and ending at $N(l)$, associated with the l-th layer,

$$F_{(0)} = [1,1,...,1]^T \tag{4.54}$$

$$F_{(l)} = \left[f_{1(l)}\left(\sum_{i(l-1)=1}^{N(l-1)} x_{i(l-1)} w_{i(l-1)1(l)} \right), f_{2(l)}\left(\sum_{i(l-1)=1}^{N(l-1)} x_{i(l-1)} w_{i(l-1)2(l)} \right), ..., f_{N(l)}\left(\sum_{i(l-1)=1}^{N(l-1)} x_{i(l-1)} w_{i(l-1)N(l)} \right) \right]^T$$

is a vector of the activation functions of the neurons associated with the l-th layer of the neural network

$\overline{W}_{(l)}$
denotes the aggregated initial weights connecting the neurons of the $(l-1)$-st layer with the l-th layer neurons.
The β token obtains now the following characteristic in place m_2

$$y(\beta) = \langle NN1, 0, E_{\max} \rangle . \tag{4.55}$$

Then, we will consider the transition Z_3, in which the new tokens of γ-type are introduced. The token γ_p, $p = 1,2...,P$, where p is the number of the training pattern, enters the place n_1 with the initial characteristic

$$y(\gamma_p) = \langle X_p(0), D_p, p \rangle \tag{4.56}$$

where $X_p(0) = [x_{p1}, x_{p2},...,x_{pN(0)}]^T$ - the input vector of the neural network, $D_p = [d_{p1}, d_{p2},...,d_{pN(0)}]^T$ - the vector of desired network outputs.

After the pattern p is applied to the network inputs as $X_p(0)$, the outputs of all layers are calculated sequentially layer by layer. The transition Z_3 describes the process of signal propagation within the neural network

$Z_3 = \langle \{\dddot{X}_4, \dddot{X}_5, \dddot{X}_9, m_2, m_7, n_1\}, \{\dddot{X}_5, \dddot{X}_6, m_3, n_2\},$

	$\dddot{X}_5$	$\dddot{X}_6$	m_3	n_2
$\dddot{X}_4$	$V_{4,5}$	$V_{4,6}$	*false*	*false*
$\dddot{X}_5$	$V_{5,5}$	$V_{5,6}$	*false*	*false*
$\dddot{X}_9$	$V_{9,5}$	$V_{9,6}$	*false*	*false*
m_2	*false*	*false*	*true*	*false*
m_7	*false*	*false*	*true*	*false*
n_1	*false*	*false*	*false*	*true*

,

$$\wedge(\vee(\dddot{X}_4, \dddot{X}_5, \dddot{X}_9), (m_2, m_7), n_1)\rangle \quad (4.57)$$

where

$V_{4,5} = V_{5,5} = V_{9,5} =$ "the previous layer does not have defined outputs",

$V_{4,6} = V_{5,6} = V_{9,6} = \neg V_{4,5}$,

$V_{1,2} =$ "all layers' outputs have assigned values for the current pattern".

In the place $\dddot{X}_5$ the tokens of α-type, $\alpha_{(l)}$, $l = 0,1,2,...,L$, obtain the new characteristics as follows

$$y(\alpha_{(l)}) = \langle NN1, l, [1, N(l)], F_{(l)}, W_{(l)}, X_{(l)}\rangle \quad (4.58)$$

where $X_{(l)} = [x_{1(l)}, x_{2(l)},..., x_{N(l)}]^T$, $l = 1,2,...,L$, is the vector of outputs of neurons associated with the l-th layer, related to the nominal weights $W_{(l)}$, $l = 1,2,...,L$.

In the place $\dddot{X}_6$ there are tokens with the following characteristics, which contain calculated neuron outputs for the pattern p

$$y(\alpha_{(l)}) = \langle NN1, l, [1, N(l)], F_{(l)}, \overline{W}_{(l)}, \overline{X}_{p(l)}\rangle \quad (4.59)$$

calculated for the nominal values of the weights $\overline{W}_{(l)}$ and states $\overline{X}_{(l)}$, $l = 1,2,...,L$.

In the place m_3 the token β preserves its characteristic as $y(\beta) = \langle NN1, 0, E_{max}\rangle$, and in the place n_2 the token γ also does not change its characteristic and remains as $y(\gamma_p) = \langle X_p(0), D_p, p\rangle$.

The next transition Z_4 describes the first stage of the estimation and weight adjustment process, which is related to the performance index computation, and has the following form

$$Z_4 = \langle \{\dddot{X}_6, m_3\}, \{\dddot{X}_7, m_4\}, \begin{array}{c|cc} & \dddot{X}_7 & m_4 \\ \hline \dddot{X}_6 & true & false \\ m_3 & false & true \end{array}, \wedge(\dddot{X}_6, m_3)\rangle . \quad (4.60)$$

As a result of computations performed within the transition Z_4, the token β obtains the new value of performance index in the place m_4,

$$y(\beta) = \langle NN1, E', E_{\max}\rangle \quad (4.61)$$

where

$$E' = E + \frac{1}{2}\sum_{j(L)=1}^{N(L)} (d_{pj(L)} - x_{pj(L)})^2 = \\ pr_2\langle NN1, E, E_{\max}\rangle + \frac{1}{2}\sum_{j(L)=1}^{N(L)} (d_{pj(L)} - x_{pj(L)})^2 \quad (4.62)$$

In the place $\dddot{X}_7$ the tokens of α -type do not change their characteristics.

In the next transition

$$Z_5 = \langle \{\dddot{X}_7, m_4\}, \{\dddot{X}_8, m_5\}, \begin{array}{c|cc} & \dddot{X}_8 & m_5 \\ \hline \dddot{X}_7 & true & false \\ m_4 & false & true \end{array}, \wedge(\dddot{X}_7, m_4)\rangle \quad (4.63)$$

the delta factors, described in Sect. 4.3, are computed in the following way

$$\delta_{pj(L)} = -\frac{\partial E_p}{\partial net_{pj(L)}} = \left(d_p - x_{pi(L-1)}\right) f'_p\left(net_{pj(L)}\right) \text{ for } j(L) = 1,2,\ldots,N(L) \quad (4.64)$$

$$\delta_{pj(l)} = f'_p\left(net_{pj(l)}\right) \sum_{k(l+1)=1}^{N(l+1)} \delta_{pk(l+1)}\, w_{pj(l)k(l+1)} \quad (4.65)$$

for $j(l) = 1,2,\ldots,N(l)$, $i(l-1) = 1,2,\ldots,N(l-1)$, and for each layer, $l = 1,2,\ldots,L$, we can write

$$\Delta_{p(l)} = \left[\delta_{p1(l)}, \delta_{p2(l)}, \ldots, \delta_{pN(l)}\right]^T .$$

The tokens of α -type obtain, in the place $\dddot{X}_8$, the following characteristics

$$y\left(\alpha_{(l)}\right) = \langle NN1, l, \left[1, N(l)\right], F_{(l)}, \overline{W}_{(l)}, \overline{X}_{p(l)}, \Delta_{p(l)}\rangle . \quad (4.66)$$

In the place m_5 the token of β -type does not change its characteristic.

The next transition Z_6, describing the process of weight adjustment, has the form

$$Z_6 = \langle \{\dddot{X}_8, m_5\}, \{\dddot{X}_9, \dddot{X}_{10}, m_6, m_7\},$$

	$\dddot{X}_9$	$\dddot{X}_{10}$	m_6	m_7
$\dddot{X}_8$	$V_{8,9}$	$V_{8,10}$	$false$	$false$
m_5	$false$	$false$	$V_{5,6}$	$V_{5,7}$

$$\wedge(\dddot{X}_8, m_5)\rangle \wedge (\dddot{X}_8, m_5)\rangle \tag{4.67}$$

where

$V_{8,9} =$ "there are still unused patterns",

$V_{8,10} = \neg V_{8,9}$,

$V_{5,6} =$ "if the performance index is below the given threshold $E_{\max}$",

$V_{5,7} = \neg V_{5,6}$.

In the place $\dddot{X}_9$ the α -type tokens obtain the new characteristic

$$y(\alpha_{(l)}) = \langle NN1,\, l,\, [1, N(l)],\, F_{(l)}, W'_{(l)} \rangle \tag{4.68}$$

with updated weight connections

$$W'_{(l)} = \left[w'_{1(l)},\, w'_{2(l)}, \ldots,\, w'_{N(l)}\right]^T$$

where

$$w'_{i(l)} = \left[w'_{i(l-1)1(l)},\, w'_{i(l-1)2(l)}, \ldots,\, w'_{i(l-1)N(l)}\right]^T$$

are calculated in the following way

$$\Delta w_{pi(L-1)j(L)} = w'_{pi(L-1)j(L)} - \overline{w}_{pi(L-1)j(L)} = \eta\, \delta_{pj(L)}\, x_{pi(L-1)}$$

for $i(L-1) = 1, 2, \ldots, N(L-1)$, $j(L) = 1, 2, \ldots, N(L)$,

$$\Delta w_{pi(l-1)j(l)} = w'_{pi(l-1)j(l)} - \overline{w}_{pi(l-1)j(l)} = \eta\, f'_p\left(net_{j(l)}\right) x_{pi(l-1)} \sum_{k(l+1)=1}^{N(l+1)} \delta_{pk(l+1}\, w_{pj(l)k(l+1)} \tag{4.69}$$

for $i(l-1) = 1, 2, \ldots, N(l-1)$, $j(l) = 1, 2, \ldots, N(l)$, and for $l = 1, 2, \ldots, L-1$,

and replace $\overline{W}[0, L-1] = W'[0, L-1]$, $\overline{X}[1, L] = X'[1, L]$.

In the place m_7 the β token obtains the characteristic $y(\beta) = \langle NN1, E, E_{max} \rangle$, which is not final.

The final values of the weights satisfying the predefined stop condition are denoted by $W^*_{(l)} = pr_5 \langle NN1, l, [1, N(l)], F_{(l)}, W'_{(l)} \rangle$, where the characteristics of the α -type tokens in the place $\dddot{X}_{10}$ are described by

$$y(\alpha_{(l)}) = \langle NN1, l, [1, N(l)], F_{(l)}, W'_{(l)} \rangle \tag{4.70}$$

and the β token characteristic in the place m_6 is described by

$$y(\beta) = \langle NN1, E', E_{max} \rangle \tag{4.71}$$

while the final value of the performance index is equal $E^* = pr_2 \langle NN1, E', E_{max} \rangle$.

The here developed generalized net representation of the backpropagation algorithm describes the main features of the gradient descent based learning algorithms. This representation allows for modifying and testing other algorithms by changing a relatively small portion of the generalized net formal description.

The generalized net methodology has been applied to describe functioning of other types of neural netwoks, see e.g. Krawczak, Atanassov and Sotirov (2010), Sotirov (2003, 2005), Sotirov and Krawczak (2003, 2006, 2008a, 2008b), Sotirov, Krawczak and Kodogiannis (2006, 2007).

Chapter 5
Learning as a Control Process

5.1 Introduction

The commonly used algorithm for the multilayer neural networks learning, the backpropagation algorithm described in the Chap. 4, is a gradient descent method for searching minimum of a performance index of learning. The performance index, being the measure of neural network learning quality, is a multimodal function. Application of this kind of algorithms causes frequent stopping at a local minimum. Various modifications of this algorithm still cannot avoid local minimal points. Until now, in practice, the only way of trying to find the near global optimum solution is to perform computation several times with different initial weight values and then to choose the best solution.

The backpropagation algorithm does not use the special layered structure of the multilayer networks. In this chapter we propose a new global algorithm for neural networks learning. The algorithm is based on the dynamic programming principle introduced by Bellman in the early 1950s (Bellman 1972, Bertsekas 1995), and allows, at least theoretically, for finding of the global minimum of the learning error. The learning of a multilayer neural network is considered as a special case of the multistage optimal control problem, first proposed by Krawczak and Mizukami (1994), and developed by Krawczak (e.g. 1995a, 1995b, 1999b, 2000a, 2001a, 2001b, 2004d, 2004g, 2005b, 2006a). The gist of the new algorithm for learning of multilayer neural networks consists of aggregating neurons within separate layers and then considering such a system as a particular multistage optimal control problem. Thus, layers become stages, while weights - controls. The problem of optimal weight adjustment is converted into a problem of optimal control.

The multistage optimal control problem can be solved by application of the dynamic programming (Bryson and Ho 1969, Cruz (1977, Roitenberg 1978, Luenberger 1984). For the new algorithm the return functions for each layer are defined, and minimization of these functions is performed layer by layer, starting from the last layer. This approach gives a real possibility of performing global optimisation. There are obstacles to the application of dynamic programming; one

M. Krawczak: *Multilayer Neural Networks*, SCI 478, pp. 95–121.
DOI: 10.1007/978-3-319-00248-4_5

is the *curse of dimensionality* – the computational burden, and the second is the memory requirement, growing exponentially with the state and control dimensionality. Fortunately, there is a way to avoid this kind of difficulties by introducing some approximation of the return functions, see Jacobson and Mayne (1979) or Yakowitz Rutherford (1984). They proposed a method to approximate the return function by considering the second-order terms in the Taylor expansion of the functions. It seems that there is possibility using the first-order method only but with application of the conjugate gradient algorithm, which converges to the inverse of the proper Hessian matrix.

In some sense, it is an application of the idea of the *neuro-dynamic programming* for the neural network learning, process introduced by Bertsekas and Tsitsiklis (1996). The term "neuro" is equivalent in this context to any kind of function approximation.

5.2 Multistage Neural Systems

Let us assume each neuron is given by the following expressions

$$x_{pj(l)} = f\left(net_{pj(l)}\right) \tag{5.1}$$

$$net_{pj(l)} = \sum_{i(l-1)=1}^{N(l-1)} w_{i(l-1)j(l)}\, x_{pi(l-1)} \tag{5.2}$$

where $x_{pj(l)}$ is the scalar output of the j-th neuron, $j(l)=1,2,...,N(l)$, situated within the layer l, $l=1,2,...,L$, the index $p=1,2,...,P$ indicates the number of a pattern, and $f\left(net_{pj(l)}\right)$ is the differentiable activation function of the neuron $j(l)$, while $net_{pj(l)}$ is the input to the neuron $j(l)$ coming from the layer $(l-1)$. The notation used is a little bit different than that used in the previous chapter in order to emphasize the stage-wise nature of the network.

Fig. 5.1 shows a multilayer neural network with distinct layers. Now, let us aggregate neurons situated within each layer l, $l=0,1,2,...,L$, in a way described by the following expressions

$$X(l) = \left[x_{1(l)}, x_{2(l)}, ..., x_{N(l)}\right]^T \quad \text{for} \quad l=0,1,2,...,L \tag{5.3}$$

$$W(l-1) = \left[w_{1(l-1)}, w_{2(l-1)}, ..., w_{N(l-1)}\right]^T \tag{5.4}$$

where $w_{j(l-1)} = \left[w_{j1(l-1)}, w_{j2(l-1)}, ..., w_{jN(l-1)}\right]^T$, for $l=1,2,...,L$, $j(l-1)=1,2,...,N(l)$.

Using Equ. 5.3 and 5.4, we can rewrite Equ. 5.1 for the whole layer (stage) in the form

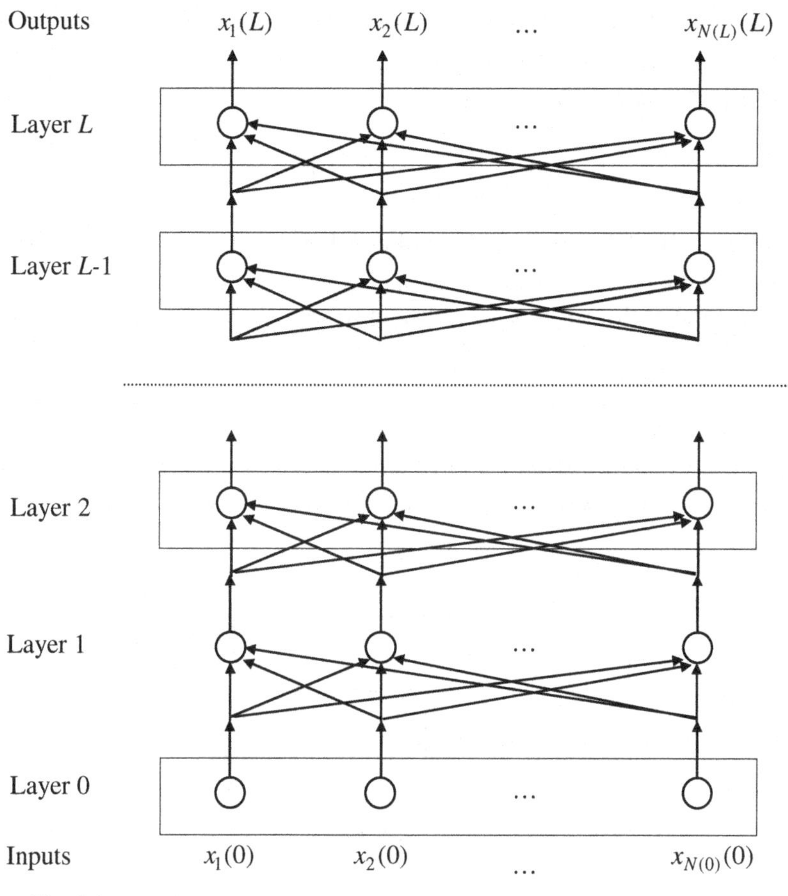

Fig. 5.1 A multilayer neural network with arranged neurons within each layer

$$X(l) = F(W(l-1),\ X(l-1)) \quad \text{for} \quad l = 1, 2, ..., L \tag{5.5}$$

where $X(l)$ denotes the aggregated output of the layer l, while $W(l-1)$ denotes the aggregated weights connecting the l-th layer with the $(l-1)$-st layer, and $X(l-1)$ is the aggregated output of the $(l-1)$-st layer.

Equ. 5.5 expresses the dynamics of the multistage system depicted in Fig. 5.2. The system is assumed to have L stages, and the evolution of the system's state through these stages, $X(0), X(1), ..., X(L)$, is governed by the equation, similar to (5.1), of the form

$$X(l+1) = F(W(l),\ X(l)) \quad \text{for} \quad l = 0, 1, ..., L-1 \tag{5.6}$$

where $F(l), l = 0,1,2,..., L,$ is a $N(l)$-dimensional vector of functions built of the separate activation functions of the neurons situated within the l-th layer. In the system theory nomenclature it is said that $X(l+1)$ denotes the output of the system in the $(l+1)$-st stage, while $W(l)$ and $X(l)$ denote the control and the input to the system associated with the $(l+1)$-st stage, respectively.

The performance index is denoted by E

$$E = \frac{1}{2}\sum_{p=1}^{P}\left\|D_p - X_p(L)\right\|^2 \tag{5.7}$$

where D_p, $p = 1,2,...,P$, is a $N(L)$ dimensional vector of the desired network outputs, while P is the number of training patterns.

This form of the performance index, in which only the output stage of the system is involved, is said in the control terminology to be in Mayer form (Sethi and Thompson, 1981).

Under the definitions (5.6) and (5.7), it is possible to define the problem of weight adjustment as an optimisation problem in a precise manner, namely the problem is to find the sequence of controls $W(l)$, $l = 0,1,...,L-1$, that minimize the performance index (5.7) subject to the state transition equation (5.6).

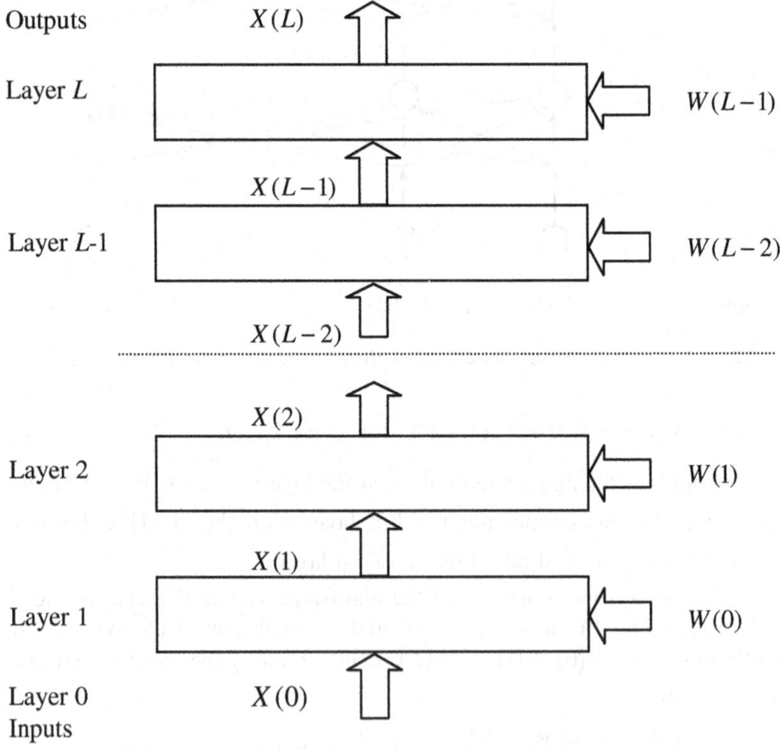

Fig. 5.2 A multilayer neural network as a multistage system

5.3 Dynamic Programming Solution

The dynamic programming solution is based on two principles that are strictly related to the structure of the problem, (Bertsekas 1995), in our case the learning problem of multilayer neural networks, (Saratchandran 1991, Krawczak and Mizukami 1994, Krawczak 1995a, 1999a, 2000a, 2001a, 2002b, 2002c).

There are two fundamental principles determining the multistage systems; one is *the principle of causality* and second - *the principle of optimality*.

The first principle states: the state $X(l)$ and the sequence of controls

$$W[l, r-1]=[W(l), W(l+1),..., W(r-1)]$$

uniquely determine the state $X(r)$, it follows directly from (5.6). According to this principle there exists a transition function $G(X(l), W[l, r-1])$, which describes the state

$$X(r)=G(X(l), W[l, r-1], l, r). \tag{5.8}$$

In this way the initial state $X(0)$ (the inputs to the neural network) and the control sequence $W[0, L-1]$ uniquely determine the sequence of states

$$X[1, L]=[X(1), X(2),..., X(L)]$$

and the performance index (5.7) can be defined as a function of the initial state $X(0)$ - and of the sequence of controls $W[0, L-1]$.

It assures that there exists a function $V(X(0), W[0, L-1])$ that determines the value of the performance index in the form

$$E=V(X(0), W[0, L-1]). \tag{5.9}$$

The principle of optimality allows for spreading of the performance index into two parts

$$E(X(0), W[0, L-1])=E_1(X(0), W[0, l-1])+E_2(X(l), W[l, L-1]). \tag{5.10}$$

The first term in (5.10) is a response of the system to the sequence of controls $W^*[0, l-1]$ that minimises E_1, the second term completes the optimisation process from the state l to the state L - due to the application of the controls $W^*[l, L-1]$, which minimize E_2.

These two principles due to Bellman determine the dynamic programming methodology. The consequence of these principles is *the principle of optimal feedback control*, which requires the optimal control at any stage to be a function of the state at this stage. Within the optimal control theory, this dependence is defined as *the optimal feedback control* which states that at the stage l, control $W(l)$ may be expressed as a function of the state $X(l)$

$$W^*(l)=W^*(X(l), l). \tag{5.11}$$

From the optimal feedback control principle (Dyer and McReynolds 1970) it follows that there exists *the return function*, which in the neural networks case has the form (Krawczak 2000a)

$$V(X(l), W[l, L-1]) = \frac{1}{2}\sum_{p=1}^{P}\left\|D_p - X_p(L)\right\|^2 \tag{5.12}$$

the control $W[l, L-1]$ being chosen in such a way in order to minimize the function (5.12).

For the optimal value of control $W^*[l, L-1]$ the return function becomes the optimal return function described by

$$V\left(X(0), W^*[0, L-1]\right) = \min_{W(0), W(1), \ldots, W(L-1)} E(X(L)). \tag{5.13}$$

For large-scale systems, like neural networks, optimisation in (5.13), i.e. finding of the optimal controls $W^*[0, L-1]$ is a very troublesome problem. The minimization in (5.13), according to the principle of optimality (Bertsekas 1995), can be treated as a stage-by-stage process. This minimization process for the whole network can be written as

$$V\left(X(0), W^*[0, L-1]\right) = \min_{W(0), W(1), \ldots, W(L-1)} E(X(L)) \tag{5.14}$$

or due to the principle of optimality (5.10)

$$\begin{aligned} &V\left(X(0), W^*[0, L-1]\right) \\ &= \min_{W(0)}\left(\min_{W(1)}\left(\ldots\left(\min_{W(L-1)} E(X(L))\right)\right)\right) \\ &= \min_{W(0)}\left(\min_{W(1)}\left(\ldots\left(\min_{W(L-1)} V(X(L))\right)\right)\right) \end{aligned} \tag{5.15}$$

where $W[0, L-1]$ is the aggregated control (the weights) associated with layers $0, 1, 2, \ldots, L-1$, while $W^*[0, L-1]$ is the optimal control (the optimal weights). The process of minimizing of Equ. 5.15 can be performed in a recursive way and can be rewritten in the following form:

for the last L-th layer

$$V(X(L)) = E(X(L))$$

for the (*L*-1)-st layer

$$V(X(L-1), W^*[L-1]) = \min_{W(L-1)} V(X(L), V(L-1))$$

...

for the *l*-th layer

$$V(X(l), W^*[l]) = \min_{W(l)} V(X(l+1), V(L-1))$$

...

for the 0-th layer

$$V(X(0), W^*[0, L-1]) = \min_{W(0)} V(X(1), W(0), W^*[1, L-1]). \tag{5.16}$$

According to Equ. 5.16, the optimisation process runs backwards starting from the output stage (the output layer) and ending at the 0-th stage (the input layer). Any stage (layer) can be described by a transition function (5.6)

$$X(l+1) = F[W(l), X(l)] \quad \text{for} \quad l = 0, 1, \ldots, L-1$$

which expresses the output $X(l+1)$of the $(l+1)$-st layer as a function of weights $W(l)$ and the output of the previous layer $X(l)$. Substituting (5.15) into (5.16), for $l = L-1$ we get the following backward transition equation

$$V(X(L-1), W^*(L-1)) = \min_{W(L-1)} E(F(X(L-1), W(L-1))). \tag{5.17}$$

Minimization of (5.17) with respect to $W(L-1)$ subject to $V(X(L)) = E(X(L))$, will give the optimal values of the controls (weights) $W^*(L-1)$ for the $(L-1)$-st stage (layer). For the $(L-2)$-nd stage (layer) the optimisation process looks like

$$V(X(L-2), W^*(L-2)) = \min_{W(L-2)} V(F(X(L-2), W(L-2), W^*(L-1))). \tag{5.18}$$

For any stage (layer) *l* the optimisation can be noted in the following form

$$V(X(l), W^*(l)) = \min_{W(l)} V(F(X(l), W(l), W^*(l+1))). \tag{5.19}$$

The main feature of the above equation is that the return function related to the *l*-th layer describes the learning error just transformed to the *l*-th layer. The optimal values of the controls $W^*(l)$ for any layer l, $l = 0, 1, 2, \ldots, L-1$ are obtained by minimization of the transformed learning error

$$V(X(l), W(l), W^*[l+1, L-1])$$

of the $(l+1)$-st stage which is expressed in terms of $X(l)$ and $W(l)$.

The return function $V(X(l), W(l), W^*[l+1, L-1])$ can be obtained as a sequence, calculated in a backward manner,

$$V(X(l), W^*[l, L-1]) = V(X(l+1), W^*[l+1, L-1]) \tag{5.20}$$

and at the last stage

$$V(X(L), W^*[L-1, L-1]) = \frac{1}{2}\sum_{p=1}^{P} \|D_p - X_p(L)\|^2 . \tag{5.21}$$

The recursive relation (5.20) is valid for any arbitrary control $W[l, L-1]$ (Dyer and McReynolds 1970), not only for the optimal

$$V(X(l), W[l, L-1]) = V(X(l+1), W[l+1, L-1]) \tag{5.22}$$

Minimization of the return function for the stage L can be performed in different ways. For more about the different optimisation methods, which can be applied to (5.17), see Bertsekas and Tsitsiklis (1996).

Generally, it is very difficult to find the solution of the dynamic programming equations. The only way to avoid the so-called *curse of dimensionality* is to apply some approximation technique for the return functions, in order to find the approximate solutions of the problem.

5.4 Return Function Approximations

In this section we consider a class of approximation, which is based on the Taylor expansion of the return functions. This class is sometimes called *the differential dynamic programming* (Yakowitz and Rutherford 1984). The term "differential dynamic programming" refers to nonlinear programming procedures based on dynamic programming. The idea of this kind of optimisation procedures was mentioned by Bellman and Dreyfus (1962), and then developed by Mayne (1966), Dyer and McReynolds (1970), Jacobson and Mayne (1970), Ohno (1978), Larson and Korsak (1970). For the learning process of the neural networks, this methodology was introduced by the present author (Krawczak 1999b, 2000a, 2000b, 2001a, 2001b).

The differential dynamic programming method is a successive approximation technique. The procedure is initiated with some nonoptimal control $\overline{W}[0, L-1]$, called *the nominal control*, which generates the *nominal trajectory* $\overline{X}[0, L]$ through the recursive formula (5.6). Within each iteration, a successor control $W'[0, L-1]$ is determined, which in result generates the performance index $E(W')$ of a lower value than $E(\overline{W})$. For the differentiable return functions $V(X, W)$ it is possible to derive a function $V(\overline{X}, \overline{W})$ consisting of a linear or quadratic part of the Taylor series expansion $V(X', W')$ of $V(X, W)$, the Taylor expansion being done about the nominal trajectory and control.

Respectively, we deal with the *first order differential dynamic programming* and the *second order differential dynamic programming*.

5.4.1 First Order Differential Dynamic Programming

The method is based on the first order expansion of the return function $V(X(0), W[0, L-1])$ about some nominal control variable sequence $\overline{W}[0, L-1]$, namely

$$\begin{aligned} & V(X'(0), W'[0, L-1]) \\ & = V(\overline{X}(0), \overline{W}[0, L-1]) + \frac{\partial V(\overline{X}(0), \overline{W}[0, L-1])}{\partial W[0, L-1]} \delta W[0, L-1]. \end{aligned} \tag{5.23}$$

The variations in the control (weight) variables $\delta W = W' - \overline{W}$ must be small enough to ensure the validity of the expansion. Choosing $\delta W = W' - \overline{W}$ as

$$\delta W[0, L-1] = \eta \left[\frac{\partial V(\overline{X}(0), \overline{W}[0, L-1])}{\partial W[0, L-1]} \right]^T \tag{5.24}$$

where η is some positive constant (in the neural network field called the learning parameter), we obtain the return function

$$V(X'(0), W'[0, L-1])$$

smaller than

$$V(\overline{X}(0), \overline{W}[0, L-1]).$$

Instead of consideration of the return function for the whole system $V(X(0), W[0, L-1])$, the backward properties (5.22) can be applied for the approximation of the return functions related to each stage.

For any $i < l$ the gradient of the return function

$$V(X(0), W[0, L-1])$$

with respect to the control $W(X(l), l)$ is equal to the gradient of

$$V(X(l), W[l, L-1])$$

that can be written as

$$\frac{\partial V(X(0), W[0, L-1])}{\partial W(X(l), l)} = \frac{\partial V(X(l), W[l, L-1])}{\partial W(X(l), l)}. \tag{5.25}$$

This property is obvious because control $W(l)$ cannot have any influence on the states $X(i)$ for $i < l$.

Let us differentiate the function $V(X(l), W[l, L-1])$ with respect to $W[l, L-1]$, where $W[l, L-1]$ is a function of the state $X(l)$, as follows

$$\frac{\partial V(X(l), W[l, L-1])}{\partial W(l)} = \left[\frac{\partial F(X(l), W(l))}{\partial W(l)}\right]^T \frac{\partial V(X(l+1), W[l+1, L-1])}{\partial X(l+1)}. \tag{5.26}$$

By considering the partial derivatives of the return function, described by Equ. 5.22, with respect to the state $X(l)$

$$V(X(l), W(l)) = V(X(l+1), W(l+1))$$

we obtain

$$\frac{\partial V(X(L))}{\partial X(l)} = -\sum_{p=1}^{P} \left\| D_p - X_p(L) \right\|$$

$$\frac{\partial V(X(l), W(l))}{\partial X(l)} = \frac{\partial F(X(l), W(l))}{\partial X(l)} \frac{\partial V(X(l+1), W(l+1))}{\partial X(l+1)}. \tag{5.27}$$

Equ. 5.26 and 5.27 can be rewritten in a shorter form

$$V_W(l) = F_W^T(l) V_X(l+1) F_W(l) \tag{5.26a}$$

$$V_X(L) = -\sum_{p=1}^{P} \left\| D_p - X_p(L) \right\|$$

$$V_X(l) = F_X(l) V_X(l+1). \tag{5.27a}$$

Using Equ. 5.26 and 5.27, the gradients of the return functions, required for Equ. 5.23 in order to derive the first order approximation, can be computed as a sequence of the equations performed from the last stage to the inputs.

The first order differential dynamic programming algorithm can be formulated in the following steps:

1. Initialise weights: $W(l)=\overline{W}(l), \quad l=0,1,2,...,L-1$

2. Set: $E=0$

3. Set: $p=0$ (p denotes the index of the pattern's number)

4. Submit a pattern: $(X_p(0), D_p), \quad p=p+1$

5. Compute the layers' outputs: $X(l)=\overline{X}(l), \quad l=1,2,..., L$ from the system equation (5.6) and the nominal weights

6. Compute the performance index of learning:

$$E=E+\frac{1}{2}\left\|D_p-\overline{X}_p(L)\right\|^2 \tag{5.28}$$

7. Compute the partial derivatives:

$V_X(l)=V_X(\overline{X}(l), \overline{W}[l, L-1])$ for $l=L, L-1,..., 0$ from

$$V_X(L)=-\left\|D_p-X_p(L)\right\| \tag{5.29}$$

$$\overline{V}_X(l)=F_X(l)\overline{V}_X(l+1) \tag{5.30}$$

8. Compute the gradient of the return function with respect to the weights for each level:

$$\frac{\partial V(\overline{X}(l), \overline{W}[l, L-1])}{\partial \overline{W}(l)} = \left[\frac{\partial F(\overline{X}(l), \overline{W}(l))}{\partial \overline{W}(l)}\right]^T \frac{\partial V(\overline{X}(l+1), \overline{W}[l+1, L-1])}{\partial \overline{X}(l+1)} \tag{5.31}$$

for $l=L-1,...,0$

9. Choose the learning parameter $\eta>0$ and compute the new weight values

$$W'(l)=\overline{W}(l)-\eta\left[\frac{\partial V(\overline{X}(l), \overline{W}[l, L-1])}{\partial \overline{W}(l)}\right]^T \tag{5.32}$$

10. If $p=P$ then go to Step 3 else go to Step 11

11. Set

$$\overline{W}[0, L-1]=W'[0, L-1]$$

$$\overline{X}[1, L]=X'[1, L]$$

12. If $E < E_{MAX}$ then go to Step 13 else go to Step 3
13. STOP.

5.4.2 First Order Differential Dynamic Programming versus Backpropagation

After having completed the description of the above algorithm, let us demonstrate the relationships between the first order differential dynamic programming algorithm and the backpropagation algorithm.

Let us consider the algorithm using a little bit different notations, which emphasises the components of the vectors and matrices.

The performance index of learning, called the learning error, to be minimized by optimal adjustment of the weights can be described as follows:

$$E = \frac{1}{2}\sum_{p=1}^{P}\sum_{j(L)=1}^{N(L)}\left(d_{pj(L)} - x_{pj(L)}\right)^2 = \frac{1}{2}\sum_{p=1}^{P}\sum_{j(L)=1}^{N(L)}\left(d_{pj(L)} - x_{pj(L)}\right)^2 \tag{5.33}$$

where p denotes an index of patterns, $j(L)$ denotes an index of output, $d_{pj(L)}$ is the desired output of the network (also called the target or pattern), and $x_{pj(L)}$ denotes the actual output of the network. We will use also the notation introduced at the beginning of this chapter, namely

$$X(l) = \left[x_{1(l)}, x_{2(l)}, \ldots, x_{N(l)}\right]^T \quad \text{for} \quad l = 0,1,2,\ldots,L$$

$$W(l-1) = \left[w_{1(l-1)}, w_{2(l-1)}, \ldots, w_{N(l-1)}\right]^T$$

$$w_{j(l-1)} = \left[w_{j1(l-1)}, w_{j2(l-1)}, \ldots, w_{jN(l)}\right]^T$$

for $l = 1,2,\ldots,L$, $j(l) = 1,2,\ldots,N(l)$.

The return function $V(X(l))$, defined as the minimum error obtained by using the optimal weight values in layers from l to L, can be rewritten for the last layer as follows

$$V(X(L)) = V(X(L)) \tag{5.34}$$

while for the $(L-1)$-st layer as

$$V(X(L-1)) = \min_{W(L-1)} E(X(L)). \tag{5.35}$$

There is a simple relation between the outputs of the L-th and $(L-1)$-st layers, Equ. 5.1 and 5.2

$$x_{pj(L)} = f\left(net_{pj(L-1)}\right) \tag{5.36}$$

$$net_{pj(L)} = \sum_{i(L-1)=1}^{N(L-1)} w_{ij(L-1)}\, x_{pi(L-1)}. \tag{5.37}$$

For brevity, we will omit the subscript in the forthcoming consideration. Substituting (5.35) and (5.36) into (5.34) we obtain

$$\begin{aligned}
&V(X(L-1))\\
&= \min_{W(L-1)} V((X(L), W(L)))\\
&= \min_{W(L-1)} V(F(X(L-1), W(L-1)))\\
&= \min_{W(l)} V\left(\begin{bmatrix} \frac{1}{2}\sum_{j(L)=1}^{N(L)}\left(d_{j(L)} - x_{j(L)}\right)^2 & & 0 \\ & \ddots & \\ 0 & & \frac{1}{2}\sum_{j(L)=1}^{N(L)}\left(d_{j(L)} - x_{j(L)}\right)^2 \end{bmatrix}\right)
\end{aligned} \tag{5.38}$$

while for any $l = 1, 2, \ldots, L-1$ we have

$$\begin{aligned}
&V(X(l))\\
&= \min_{W(l)} V(F(X(l), W(l)))\\
&= \min_{W(l)} V\left(\begin{bmatrix} f\left(x_{1(l)}, net_{1(l)}\right) & \cdots & \\ \vdots & \ddots & \\ & & f\left(x_{N(l)}, net_{N(l)}\right) \end{bmatrix}\right)\\
&= \min_{W(l)} V((X(l+1), W(l+1))).
\end{aligned} \tag{5.39}$$

Now, let us consider the continuous activation functions described in the previous chapter. By rewritting Equ. 5.32 for updating a single weight $w_{i(L-1)j(L)}(L-1)$ we can obtain the following equation

$$w'_{i(L-1)j(L)} = \overline{w}_{i(L-1)j(L)} + \Delta\overline{w}_{i(L-1)j(L)} \tag{5.40}$$

where the correction term is

$$\Delta \overline{w}_{i(L-1)\,j(L)} = -\eta \left[\frac{\partial\, V\left(\overline{X}(L)\right)}{\partial\, \overline{w}_{i(L-1)\,j(L)}} \right]. \tag{5.41}$$

Let us deploy the details in the square bracket of Equ. 5.41

$$\begin{bmatrix} 0 & \cdots & 0 & \cdots & 0 \\ \vdots & & \dfrac{\partial\, f\left(x_i(L-1), net_{j(L-1)}\right)}{\partial\, w_{i(L-1)\,j(L)}} & & \vdots \\ 0 & \cdots & 0 & \cdots & 0 \end{bmatrix}$$

and we can notice that all the components are zero, except the one indicated by indices $i(L-1)\,j(L)$, thus the correction term (5.41) becomes

$$\begin{aligned} &\Delta \overline{w}_{i(L-1)\,j(L)} \\ &= -\eta \left[\frac{\partial \frac{1}{2} \sum_{j(L)=1}^{N(L)} \left(d_j - f\left(x_{i(L-1)}, net_{j(L)}\right)\right)^2}{\partial\, \overline{w}_{i(L-1)\,j(L)}} \right] \\ &= -\eta\, \frac{\partial \frac{1}{2} \sum_{j(L)=1}^{N(L)} \left(d_{j(L)} - f\left(x_{i(L-1)}, net_{j(L)}\right)\right)^2}{\partial net_{j(L)}} \frac{\partial net_{j(L)}}{\partial\, \overline{w}_{i(L-1)\,j(L)}} \\ &= \eta\ \delta_{j(L)} \frac{\partial net_{j(L)}}{\partial\, \overline{w}_{i(L-1)\,j(L)}} \\ &= \eta\ \delta_{j(L)}\, \overline{x}_{i(L-1)} \end{aligned} \tag{5.42}$$

where

$$\delta_{j(L)} = -\frac{\partial V(X(L))}{\partial net_{j(L)}} = \left(d_{j(L)} - f\left(x_{i(L-1)}, net_{j(L)}\right)\right) \frac{\partial f\left(x_{i(L-1)}, net_{j(L)}\right)}{\partial net_{j(L)}}.$$

Using a gradient search procedure for minimizing the performance index (5.38) we obtain the optimal value of

$$net^{*}_{j(L)} = \sum_{i(L-1)=1}^{N(L-1)} w^{*}_{i(L-1)j(L)}\, x_{i(L-1)} \tag{5.43}$$

and the performance index becomes

$$V(X(L-1)) = \sum_{j(L-1)=1}^{N(L-1)} \left(d_{j(L-1)} - f\left(net^*_{j(L-1)}\right)\right)^2. \tag{5.44}$$

For the $(L-2)$ -nd layer the correction term becomes

$$\begin{aligned} &\Delta \overline{w}_{i(L-2)j(L-1)} \\ &= \eta\, \delta_{j(L-1)} \frac{\partial net_{j(L-1)}}{\partial \overline{w}_{i(L-2)j(L-1)}} \\ &= \eta\, \delta_{j(L-1)} \sum_{j=1}^{N(L)} \delta_{j(L)} \overline{w}_{i(L-1)j(L)} \end{aligned} \tag{5.45}$$

where

$$\begin{aligned} \delta_{j(L-1)} &= -\frac{\partial V(X(L))}{\partial net_{j(L-1)}} \\ &= \left(d_{j(L-1)} - f\left(x_{i(L-1)}, net_{j(L)}\right)\right) \frac{f\left(x_{i(L-2)}, net_{j(L-1)}\right)}{\partial net_{j(L-1)}} \end{aligned}$$

$$\delta_{j(L)} = -\frac{\partial V(X(L))}{\partial net_{j(L-1)}} = \left(d_{j(L)} - f\left(x_{i(L-1)}, net_{j(L)}\right)\right) \frac{f\left(x_{i(L-1)}, net_{j(L)}\right)}{\partial net_{j(L)}}.$$

Updating of any weight $w_{i(l-1)j(l)}$, where the indices are ordered as follows

$$i(l-1) = 1,2,\ldots,N(l-1),\ \ j(l) = 1,2,\ldots,N(l),\ \ l = 1,2,\ldots,L$$

is expressed by

$$w'_{i(l-1)j(l)} = \overline{w}_{i(l-1)j(l)} + \Delta \overline{w}_{i(l-1)j(l)} \tag{5.46}$$

while the correction term $\Delta \overline{w}_{i(l-1)j(l)}$ for any $l = 1,2,\ldots,L-1$ has the following form

$$\begin{aligned} &\Delta \overline{w}_{i(l-1)j(l)} \\ &= \eta\, \delta_{j(l)} \sum_{j(l+1)=1}^{N(l+1)} \delta_{j(l+1)} \overline{w}_{i(l)j(l+1)} \sum_{j(l+2)=1}^{N(l+2)} \delta_{j(l+2)} \overline{w}_{i(l+1)j(l+2)} \cdots \sum_{j(L)=1}^{N(L)} \delta_{j(L)} \overline{w}_{i(L-1)j(L)} \end{aligned} \tag{5.47}$$

under the definition of the δ

$$\delta_{j(l)} = \overline{w}_{i(l-1)j(l)} \frac{f\left(x_{i(l-1)}, net_{j(l)}\right)}{\partial net_{j(l-1)}}. \tag{5.48}$$

The formulae (5.46) – (5.48) obtained describe the correction of weights in the learning process, based on the first order differential dynamic programming. The equations look exactly the same as in the standard backpropagation algorithm. However, there is an essential difference between the two algorithms. In the backpropagation algorithm, the weights are changed a little bit in order to improve the performance index slightly. In the case of the algorithm proposed in this chapter, within each layer (stage) the changes of the weights are performed until the minimum of the proper return function is reached. After that, the computation is shifted to another layer of lower index l up to the one labelled by 0.

5.4.3 Second Order Differential Dynamic Programming

The method is based on the second order expansion of the return function $V(X(0), W[0, L-1])$ about some nominal control variable sequence $\overline{W}[0, L-1]$, which generates the nominal states (trajectory) of the multistage neural system, namely

$$V(X'(0), W'[0, L-1]) = \delta W^T V_{WX}\, \delta X + \frac{1}{2} \delta W^T V_{WW}\, \delta W + V_W^T\, \delta W + V_X\, \delta X \tag{5.49}$$

where V_{XX}, V_{WW} and V_{XW} are symmetric, positive definite matrices of the proper order, while V_W and V_X are vectors of proper orders, $\delta X = X' - \overline{X}_L$ and $\delta W = W' - \overline{W}_L$ are small enough to ensure the validity of the expansion. Equ. 5.49 is a quadratic approximation of the return function. The necessary and sufficient condition for the control $W'(X, L)$ to minimize (5.49) is to fulfil the following condition

$$\left(\frac{\partial V}{\partial W}\right) = V_{WW}\, \delta W + V_{XX}\, \delta X + V_W = 0. \tag{5.50}$$

Equ. 5.50 allows us to obtain the optimal control (weight) increment

$$\delta W = \frac{1}{2} V_{WW}^{-1} \left(V_W + V_{XX}\, \delta X\right) = \alpha_0 + \beta_0\, \delta X \tag{5.51}$$

where

$$\begin{aligned} \delta W &= W'[0, L-1] - \overline{W}[0, L-1] \\ \alpha_0 &= -\frac{1}{2} V_{WW}^{-1}[0, L-1]\, V_W[0, L-1] \\ \beta_0 &= -\frac{1}{2} V_{WW}^{-1}[0, L-1]\, V_{WX}[0, L-1]. \end{aligned} \tag{5.52}$$

The above derivation has been done under the assumption that the Hessian matrix V_{WW} is nonsingular. Similarly as in the first order differential dynamic programming case, this method can be implemented in a sequential manner, starting from the last stage.

Let us first consider the last layer governed by $W(L-1)$ and $X(L-1)$, the return function has the form

$$\begin{aligned} &V\left(\overline{X}(L-1), \overline{W}(L-1)\right) \\ &= \frac{1}{2}\sum_{p=1}^{P}\left\|D_p - X_p(L)\right\|^2 \\ &= \frac{1}{2}\sum_{p=1}^{P}\left\|D_p - F_p\left(\overline{X}(L-1), \overline{W}(L-1)\right)\right\|^2 \end{aligned} \tag{5.53}$$

which allows us to calculate

$$\begin{aligned} \alpha_{L-1} &= -\frac{1}{2}V_{WW}^{-1}(L-1)V_W(L-1) \\ \beta_{L-1} &= -\frac{1}{2}V_{WW}^{-1}(L-1)V_{XW}(L-1) \end{aligned} \tag{5.54}$$

in order to find

$$\begin{aligned} W'(L-1) &= \overline{W}(L-1) + \alpha_{L-1} + \beta_{L-1}\,\delta X(L-1) \\ &= \arg\min_{W(L-1)} V\left(\overline{X}(L-1), W'(L-1)\right). \end{aligned} \tag{5.55}$$

Equ. 5.55 can be substituted into (5.53) in order to get the new and lower value of the return function

$$V\left(X'(L-1), W'(L-1)\right) = \frac{1}{2}\sum_{p=1}^{P}\left\|D_p - F_p\left(\overline{X}(L-1), W'(L-1)\right)\right\|^2. \tag{5.56}$$

Using (5.56) we can calculate the following vector and matrix

$$\begin{aligned} Q^T(L-1) &= \frac{\partial V\left(X'(L-1), W'(L-1)\right)}{\partial X(L-1)} \\ P(L-1) &= \frac{1}{2}\frac{\partial^2 V\left(X'(L-1), W'(L-1)\right)}{\partial X^2(L-1)}. \end{aligned} \tag{5.57}$$

Similarly, for any $l = L-2, L-3, \ldots, 1, 0$ we can find the quadratic approximation of the return functions, as follows

$$\begin{aligned} &V\left(X'(l), W'(l)\right) \\ &= \left(F\left(\overline{X}(l), \overline{W}(l)\right) - X'(l+1)\right)^T P(l+1)\left(F\left(\overline{X}(l), \overline{W}(l)\right) - X'(l+1)\right) \\ &+ Q^T(l+1)\left(F\left(\overline{X}(l), \overline{W}(l)\right) - X'(l+1)\right). \end{aligned} \tag{5.58}$$

After the calculation of

$$\begin{aligned} \alpha_l &= -\frac{1}{2} V_{WW}^{-1}(l) V_W(l) \\ \beta_l &= -\frac{1}{2} V_{WW}^{-1}(l) V_{WX}(l) \end{aligned} \tag{5.59}$$

it is possible to find the weights

$$W'(l) = \overline{W}(l) + \alpha_l + \beta_l \, \delta X(l) = \underset{W(l)}{\arg\min} \, V\left(\overline{X}(l), W'(l)\right). \tag{5.60}$$

Equ. 5.60 allows us to describe $V\left(\overline{X}(l), W'(l)\right)$, and then the coefficients

$$\begin{aligned} Q^T(l) &= \frac{\partial V(X'(l), W'(l))}{\partial X(l)} \\ P(l) &= \frac{1}{2} \frac{\partial^2 V(X'(l), W'(l))}{\partial X^2(l)} . \end{aligned} \tag{5.61}$$

After computing and storing all α_l, β_l for $l = 0,1,2,\ldots,L-1$ we can obtain the new states of the system (new neuron outputs) in the following way

$$X'(l) = F\left(\overline{X}(l), W'(l)\right) \text{ for } l = 1,2,\ldots,L$$

upon which the obtained weights and states become the nominal ones, as

$$\overline{X}(l) = X'(l), \quad \overline{W}(l) = W'(l)$$

which allows us to start the second order differential dynamic programming procedure from the beginning.

The above method requires computation of the inverse matrices of the Hessians. Even such computation appearing in Equ. 5.54 and 5.59, does not make a problem nowadays.

Moreover, there are several methods, which use the Hessian of the performance index, and the problem of matrix inversion can be overcome by the approximating methods (Smagt 1994), like:

- o Quasi-Newton Methods.
- o Conjugate Gradient Methods.

These methods allow computing the Hessian's inverse iteratively.

There are also different ways, which make it possible to overcome the computational problems. In this book we would like to mention the method based on the first order differential programming with the use of the neuron model parameters, and also a method known in game theory for the incremental learning process. Both methods are described in the next chapter.

5.5 Description of First Order Differential Dynamic Programming Algorithm

The generalized net representation of the backpropagation algorithm was described in Sect. 4.6. The learning algorithms based on differential dynamic programming in Sect. 5.3 and 5.4. In this section, we will consider the process of neural network training as a multistage optimal control problem, and we will use the generalized net methodology in order to describe the neural networks learning algorithm based on the first order differential dynamic programming.

The procedure for developing this new representations of neural networks based on generalized net methodology is based first of all on the books by Atanassov (1991, 1998, 2007), and by Atanassov and Aladjov (2000), as well as the papers by Krawczak et al. (2003), and the works of Krawczak (2003b, 2003e, 2004e, 2004g, 2005b).

Similarly as in Sect. 4.6 it is assumed that the basic dynamic components of the neural network are neurons. This assumption is crucial because the signals, as well as the connection weights are less important in this consideration. In such a manner all changes in the neural network structure and states are represented by changes of neuron states. The α -type tokens describing each neuron (or a group of neurons) in the neural network enter the net through the place $\dddot{X}_1$, and have the following initial characteristics

$$y\left(\alpha_{i(l)}\right) = \langle NN1,\ l,\ i(l),\ f_{i(l)}\rangle \tag{5.62}$$

for $i(l) = 1,2,\ldots,N(l)$, $l = 0,1,\ldots,L$, where

$NN1$
the neural network identifier,

i
the number of the token (neuron) associated with the l -th layer,

l
the present layer number,

$f_{1(l)}(\cdot)$
the activation function of the i -th neuron associated with the l -th layer of the neural network, where $f_{i(0)}(\cdot) = 1.0$, for $i(0) = 1,2,\ldots,N(0)$.

The generalized net representation of the first order differential dynamic programming algorithm contains seven transitions, see Fig. 5.3. It is particularly important that most of the transitions are the same as for the backpropagation algorithm (therefore, they will be reminded in a short way), and the transition Z_5 will be divided into two transitions, $Z_{5,1}$ and $Z_{5,2}$.

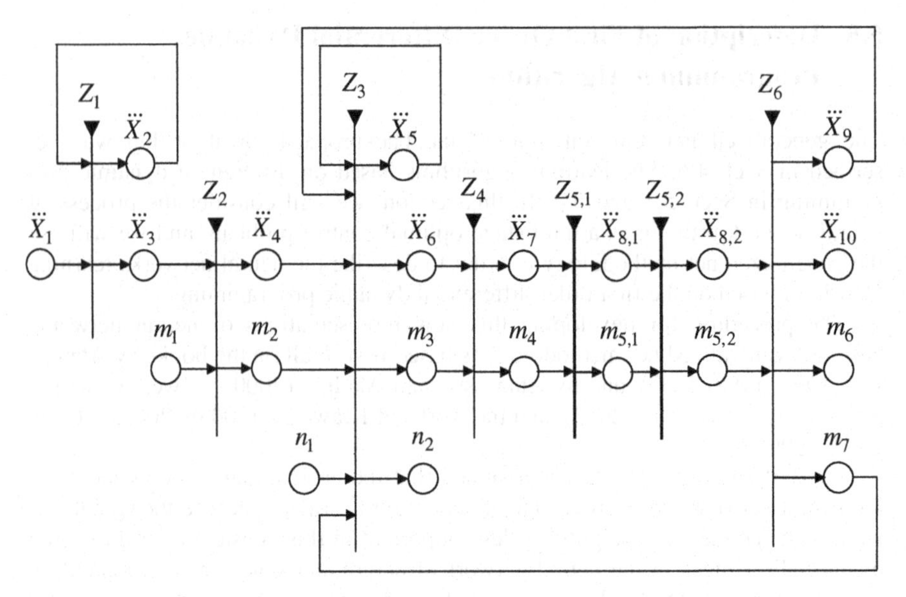

Fig. 5.3 The generalized net description of the first order differential dynamic programming algorithm

Via the transition Z_1 every token $\alpha_{i(l)}$, $i(l)=1,2,...,N(l)$, $l=0,1,...,L$, is transferred from the place $\dddot{X}_1$ to the place $\dddot{X}_2$ as well as $\dddot{X}_3$. The tokens are transferred sequentially according to the increasing indexes $i(l)=1,2,...,N(l)$ for $l=0,1,...,L$, in this way the tokens of the same level l are aggregated into a new token $\alpha_{(l)}$, representing the layer l, according the condition transitions in the transition Z_1

$$Z_1 = \langle \{\dddot{X}_1, \dddot{X}_2\}, \{\dddot{X}_2, \dddot{X}_3\}, \begin{array}{c|cc} & \dddot{X}_2 & \dddot{X}_3 \\ \hline \dddot{X}_1 & V_{1,2} & V_{1,3} \\ \dddot{X}_2 & V_{2,2} & V_{2,3} \end{array}, \vee(\dddot{X}_1, \dddot{X}_2)\rangle \tag{5.63}$$

where

$V_{1,2} = \neg V_{1,3} =$ "if there is only one token $\alpha_{i(l)}$ in the place $\dddot{X}_1$", i.e.

$$\forall\left(\alpha_{j(l)} \in K_{\dddot{X}_1}\right)\left(pr_2 Y_{\alpha_{i(l)}} \neq pr_2 Y_{\alpha_{j(l)}}, j(l) \neq i(l)\right) \tag{5.64}$$

(where $K_{\dddot{X}_1}$ is a set of all tokens entering the net from the place $\dddot{X}_1$, $i(l), j(l)=1,2,...,N(l)$)

$V_{2,2}$ = "if there is more than one token $\alpha_{i(l)}$ and $\alpha_{j(l)}$ associated with the l-th layer", i.e.,

$$\exists\left(\alpha_{i(l)} \in K_{\ddot{X}_1} \ \& \ \alpha_{j(l)} \in K_{\ddot{X}_2}\right)\left(pr_2 Y_{\alpha_{i(l)}} = pr_2 Y_{\alpha_{j(l)}}\right) \tag{5.65}$$

$V_{2,3}$ = "if all tokens $\alpha_{i(l)}$, $i(l) = 1, 2, \ldots, N(l)$, have been combined into one token" i.e.,

$$\begin{aligned} &\neg\exists\left(\alpha_{i(l)} \in K_{\ddot{X}_1} \ \& \ \alpha_{k(l)} \in K_{\ddot{X}_2}\right)\left(pr_2 Y_{\alpha_{i(l)}} = pr_2 Y_{\alpha_{k(l)}}\right) \ \& \\ &\neg\exists\left(\alpha_{i(l)} \in K_{\ddot{X}_1} \ \& \ \alpha_{j(l)} \in K_{\ddot{X}_1}\right)\left(pr_2 Y_{\alpha_{i(l)}} = pr_2 Y_{\alpha_{j(l)}}, i \neq j\right). \end{aligned} \tag{5.66}$$

The tokens associated with neurons lying within one layer are aggregated in the transition Z_1. For layer, $l = 0, 1, \ldots, N$, the characteristics of the tokens are processed in order to construct a new token $\alpha_{(l)}$ representing the whole l-th layer according to the condition (5.66). The aggregated token is transferred from the place $\dddot{X}_2$ to the place $\dddot{X}_3$, and has the following characteristic

$$y\left(\alpha_{(l)}\right) = \langle NN1,\ l,\ [1, N(l)],\ F_{(l)} \rangle \quad \text{for } l = 0, 1, 2, \ldots, L, \tag{5.67}$$

where

$NN1$
the neural network identifier,

l
the layer number,

$[1, N(l)]$
denotes $N(l)$ tokens (neurons) arranged in a sequence, starting form the first and ending at $N(l)$, associated with the l-th layer,

$F_{(0)} = [1, 1, \ldots, 1]^T$

$$F_{(l)} = \left[f_{1(l)}(\cdot), f_{2(l)}(\cdot), \ldots, f_{N(l)}(\cdot)\right]^T \tag{5.68}$$

is a vector of the activation functions of the neurons associated with the l-th layer of the neural network.

In the place $\dddot{X}_3$ we obtain L tokens, producing the neural network output.

The second transition Z_2 has the following form

$$Z_2 = \langle \{\dddot{X}_3, m_1\}, \{\dddot{X}_4, m_2\}, \begin{array}{c|cc} & \dddot{X}_4 & m_2 \\ \hline \dddot{X}_3 & true & false \\ m_1 & false & true \end{array}, \wedge\left(\dddot{X}_3, m_1\right) \rangle \tag{5.69}$$

where the performance index of the learning process is introduced, and the β-type token, which enters the input place m_1, has the following initial characteristic

$$y(\beta) = \langle NN1, E, E_{max} \rangle \tag{5.70}$$

where

$NN1$
the neural network identifier,

E
the performance index of the neural network learning,

E_{max}
the threshold value of the performance index, which must be reached.

The token $\alpha_{(l)}$, $l = 0,1,2,...,L$, related to the l-th layer, has now the characteristic, in the place $\dddot{X}_4$,

$$y(\alpha_{(l)}) = \langle NN1, l, [1, N(l)], F_{(l)}, \overline{W}_{(l)} \rangle \tag{5.71}$$

for $l = 0,1,2,...,L$, where

$NN1$
the neural network identifier,

l
the layer number,

$[1, N(l)]$
denotes $N(l)$ tokens (neurons) arranged in a sequence, starting form the first and ending at $N(l)$, associated with the l-th layer,

$$F_{(0)} = [1,1,...,1]^T, \quad F_{(l)} = [f_{1(l)}(\cdot), f_{2(l)}(\cdot), ..., f_{N(l)}(\cdot)]^T \tag{5.72}$$

is a vector of the activation functions of the neurons associated with the l-th layer

$\overline{W}_{(l)}$
denotes the aggregated initial weights connecting the neurons of the l-th layer with the $(l-1)$-st layer neurons.

In place m_2 the β token obtains now the characteristic

$$y(\beta) = \langle NN1, 0, E_{max} \rangle. \tag{5.73}$$

In the transition Z_3 the tokens γ_p, $p = 1,2...,P$, p being the number of the training pattern, enters the place n_1 with the initial characteristic

$$y(\gamma_p) = \langle X_p(0), D_p, p \rangle \qquad (5.74)$$

where

$X_p(0) = [x_{p1}, x_{p2}, \ldots, x_{pN(0)}]^T$
is the input vector of the neural network,

$D_p = [d_{p1}, d_{p2}, \ldots, d_{pN(0)}]^T$
is the vector of desired network outputs,

and for the input $X_p(0)$ the outputs of all layers are calculated.

The transition Z_3 describes the process of signal propagation in the neural network (5.76). The tokens $\alpha_{(l)}$, $l = 0,1,2,\ldots,L$, in the place $\dddot{X}_5$, obtain the new characteristics

$$y(\alpha_{(l)}) = \langle NN1, l, [1, N(l)], F_{(l)}, W_{(l)}, X_{(l)} \rangle \qquad (5.75)$$

where $X_{(l)}$, $W_{(l)}$, $l = 1,2,\ldots,L$.

$Z_3 = \langle \{\dddot{X}_4, \dddot{X}_5, \dddot{X}_9, m_2, m_7, n_1\}, \{\dddot{X}_5, \dddot{X}_6, m_3, n_2\},$

	$\dddot{X}_5$	$\dddot{X}_6$	m_3	n_2
$\dddot{X}_4$	$V_{4,5}$	$V_{4,6}$	*false*	*false*
$\dddot{X}_5$	$V_{5,5}$	$V_{5,6}$	*false*	*false*
$\dddot{X}_9$	$V_{9,5}$	$V_{9,6}$	*false*	*false*
m_2	*false*	*false*	*true*	*false*
m_7	*false*	*false*	*true*	*false*
n_1	*false*	*false*	*false*	*true*

,

$$\wedge(\vee(\dddot{X}_4, \dddot{X}_5, \dddot{X}_9), (m_2, m_7), n_1)\rangle \qquad (5.76)$$

where

$V_{4,5} = V_{5,5} = V_{9,5} =$ "previous layer does not have defined outputs"

$V_{4,6} = V_{5,6} = V_{9,6} = \neg V_{4,5}$

$V_{1,2} =$ "all layers' outputs have assigned values for the current pattern".

In the place $\dddot{X}_6$ the tokens α get the following characteristics

$$y(\alpha_{(l)}) = \langle NN1,\, l,\, [1, N(l)],\, F_{(l)}, \overline{W}_{(l)}, \overline{X}_{p(l)} \rangle \tag{5.77}$$

while the token β has the characteristic $y(\beta) = \langle NN1, 0, E_{\max} \rangle$ (in the place m_3), and the token γ has the characteristic $y(\gamma_p) = \langle X_p(0),\, D_p, p \rangle$ in the place n_2.

The transition Z_4 describes the estimation and weight adjustment, and has the form

$$Z_4 = \langle \{\dddot{X}_6, m_3\}, \{\dddot{X}_7, m_4\}, \begin{array}{c|cc} & \dddot{X}_7 & m_4 \\ \hline \dddot{X}_6 & true & false \\ m_3 & false & true \end{array}, \wedge(\dddot{X}_6, m_3) \rangle . \tag{5.78}$$

In the place m_4 the token β generates the new value of the performance index via the characteristic

$$y(\beta) = \langle NN1, E', E_{\max} \rangle \tag{5.79}$$

where

$$E' = E + \frac{1}{2} \sum_{j(L)=1}^{N(L)} (d_{pj(L)} - x_{pj(L)})^2 = pr_2 \langle NN1, E, E_{\max} \rangle^{\beta} + \frac{1}{2} \sum_{j(L)=1}^{N(L)} (d_{pj(L)} - x_{pj(L)})^2 . \tag{5.80}$$

The tokens of α -type do not change their characteristics in the place $\dddot{X}_7$.

In distinction from the generalized net representation for the classic backpropagation algorithm, described in Chap. 4, here the transition Z_5 is split in two transitions $Z_{5,1}$ and $Z_{5,2}$.

The transition $Z_{5,1}$ models the evaluation of the partial derivatives described in details in Chap. 4

$$Z_{5,1} = \langle \{\dddot{X}_7, m_4\}, \{\dddot{X}_{8,1}, m_{5,1}\}, \begin{array}{c|cc} & \dddot{X}_{8,1} & m_{5,1} \\ \hline \dddot{X}_7 & true & false \\ m_4 & false & true \end{array}, \wedge(\dddot{X}_7, m_4) \rangle . \tag{5.81}$$

The partial derivatives of the return functions $V(X(l), W[l, L-1])$ with respect to $W[l, L-1]$, where $W[l, L-1]$ is a function of the state $X(l)$, are described by (5.26) (here 5.82)

$$\frac{\partial V(X(l), W[l, L-1])}{\partial W(l)} = \left[\frac{\partial F(X(l), W(l))}{\partial W(l)} \right]^T \frac{\partial V(X(l+1), W[l+1, L-1])}{\partial X(l+1)} \tag{5.82}$$

and partial derivatives of the return functions with respect to the aggregated (within each layer) states $X(l)$ are described by (5.27) (here 5.83)

$$\frac{\partial V(X(L))}{\partial X(l)} = -\sum_{p=1}^{P} \left\| D_p - X_p(L) \right\|$$
$$\frac{\partial V(X(l), W(l))}{\partial X(l)} = \frac{\partial F(X(l), W(l))}{\partial X(l)} \frac{\partial V(X(l+1), W(l+1))}{\partial X(l+1)}. \tag{5.83}$$

For the given nominal values of the neuron states $\overline{X}(l)$ and weight connections $\overline{W}[l, L-1]$, where $l = L-1,...,0$, the equations (5.82) and (5.83) in a shorter form are as follows

$$\overline{V}_X(l) = V_X\left(\overline{X}(l), \overline{W}[l, L-1]\right) \tag{5.84}$$

$$V_X(L) = -\sum_{p=1}^{P} \left\| D_p - X_p(L) \right\|$$
$$\overline{V}_X(l) = F_X(l) \overline{V}_X(l+1) \tag{5.85}$$

for $l = L-1,...,0$.

The tokens of α-type obtain, in the place $\dddot{X}_{8,1}$, the following characteristics, which include the results of Equ. 5.83

$$y(\alpha_{(l)}) = \langle NN1,\ l,\ [1, N(l)], F_{(l)}, \overline{W}_{(l)}, \overline{X}_{p(l)}, \overline{V}_X(l) \rangle. \tag{5.86}$$

The token of β-type does not change its characteristic in the place $m_{5,1}$.

The transition $Z_{5,2}$ is devoted to the partial derivatives of the return functions $V(X(l), W[l, L-1])$ with respect to $W[l, L-1]$, which are described by (5.82), and have the following form

$$Z_{5,2} = \langle \{\dddot{X}_{8,1}, m_{5,1}\}, \{\dddot{X}_{8,2}, m_{5,2}\}, \begin{array}{c|cc} & \dddot{X}_{8,2} & m_{5,2} \\ \hline \dddot{X}_{8,1} & true & false \\ m_{5,1} & false & true \end{array}, \wedge(\dddot{X}_{8,1}, m_{5,1}) \rangle. \tag{5.87}$$

The tokens of α-type obtain, in the place $\dddot{X}_{8,2}$, the following characteristics

$$y(\alpha_{(l)}) = \langle NN1,\ l,\ [1, N(l)], F_{(l)}, \overline{W}_{(l)}, \overline{X}_{p(l)}, \overline{V}_W(l) \rangle \tag{5.88}$$

where $\overline{V}_W(l) = F_W^T(l) \overline{V}_X(l+1) F_W(l)$ for $l = L-1,...,0$, while the token β has not changed its characteristic.

The transition Z_6 describes the process of weights adjustment during the learning process and has the following form

$$Z_6 = \langle \{\dddot{X}_{8,2}, m_{5,2}\}, \{\dddot{X}_9, \dddot{X}_{10}, m_6, m_7\},$$

	$\dddot{X}_9$	$\dddot{X}_{10}$	m_6	m_7
$\dddot{X}_{8,2}$	$V_{8,9}$	$V_{8,10}$	*false*	*false*
$m_{5,2}$	*false*	*false*	$V_{5,6}$	$V_{5,7}$

$$\wedge (\dddot{X}_{8,2}, m_{5,2})\rangle \quad (5.89)$$

where

$V_{8,9} =$ "there are still unused patterns",

$V_{8,10} = \neg V_{8,9}$,

$V_{5,6} =$ "if the performance index is below the given threshold $E_{\max}$",

$V_{5,7} = \neg V_{5,6}$.

The α -type tokens obtain the new characteristics in the place $\dddot{X}_9$

$$y(\alpha_{(l)}) = \langle NN1, l, [1, N(l)], F_{(l)}, W'_{(l)} \rangle \quad (5.90)$$

with updated weight connections

$$W'_{(l)} = [w'_{1(l)}, w'_{2(l)}, ..., w'_{N(l)}]^T \quad (5.91)$$

where

$$w'_{i(l)} = [w'_{i(l-1)1(l)}, w'_{i(l-1)2(l)}, ..., w'_{i(l-1)N(l)}]^T$$

for $i(L-1) = 1, 2, ..., N(L-1)$, $j(L) = 1, 2, ..., N(L)$. The new values of the weights are calculated in the following way (for the learning parameter $\eta > 0$)

$$W'(l) = \overline{W}(l) - \eta \left[\frac{\partial v(\overline{X}(l), \overline{W}[l, L-1])}{\partial \overline{W}(l)} \right]^T \quad (5.92)$$

for $l = L, L-1, ..., 0$, and replace $\overline{W}[0, L-1] = W'[0, L-1]$, $\overline{X}[1, L] = X'[1, L]$.

In the place m_7 the β token obtains the characteristic

$$y(\beta) = \langle NN1, E, E_{\max} \rangle$$

which is not final.

The optimal values of the weights satisfying the stop condition are denoted by $W_{(l)}^{*} = pr_5 \langle NN1, l, [1, N(l)], F_{(l)}, W'_{(l)} \rangle$, where the characteristics of the α-type tokens, in the place $\dddot{X}_{10}$, is described by

$$y(\alpha_{(l)}) = \langle NN1, l, [1, N(l)], F_{(l)}, W'_{(l)} \rangle \tag{5.93}$$

while the final value of the performance index is equal $E^{*} = \langle NN1, E', E_{\max} \rangle$, and the β token characteristic in the place m_6 is described by

$$y(\beta) = \langle NN1, E', E_{\max} \rangle. \tag{5.94}$$

Chapter 6
Parameterisation of Learning

6.1 Introduction

Learning of a neural network is meant to adjust connections between layers (connections between neurons) in order to minimize the performance index of learning. For this, the backpropagation algorithm with various modifications is commonly used. At the same time, the learning process of multilayer neural networks can be considered as a particular multistage optimal control problem, described in Chap. 4. This kind of problem can be naturally treated by the dynamic programming approach (Chernousko and Lyubushin 1982, Larson and Korsak 1970) as well as (Krawczak 1994, 1995a, 1999b, 2000a, 2001b, 2002b, 2002c).

In this chapter we introduce a gain parameter into models of neurons. The value of this parameter is tacitly assumed to be 1.0 in almost all learning algorithms used. Note that setting the parameter to a small value makes the neuron model "almost linear". Thus, the learning process problem can be solved using computational tools specified for *linear-quadratic* systems optimisation, like the first order differential dynamic programming methodology described in Chap. 5. Using the continuation methodology (Krawczak 1999b, 2000b, 2000c, 2000e) the value of the gain parameter may be changed in order to reach 1.0. In fact, we can do much more, namely by considering the gain parameter as an additional control variable, the optimal value of the parameter can be found (Krawczak 2001a, 2002b). The presented methodology is based on the first order differential dynamic programming, and due to the global properties of the methodology we propose to call it *the heuristic dynamic programming*. In some sense the idea is borrowed from the simulated annealing approach, known in stochastic mechanics (Aarts and Laarhoven 1987, Kirkpatrick et al. 1983, Brooks and Morgan 1995).

In this chapter we will also describe the method of conversion of the multilayer neural networks learning problem into the iterative minimax problem. The method uses the methodology of game theory as well as the tools of multiobjective optimisation introduced by Krawczak (1997b, 1998) and allows to consider the neural network learning process as a multiobjective optimisation problem, where each pair of the training examples is associated with a partial performance index

M. Krawczak: *Multilayer Neural Networks*, SCI 478, pp. 123–144.
DOI: 10.1007/978-3-319-00248-4_6

(as a separate objective function). The methodology can be used for establishing a new paradigm of learning that we will call *the updating learning*.

6.2 Neuron Models with Parameters

Let us consider an artificial neuron model which is a nonlinear processing unit performing the operation $f(net)$ by means of the activation function, where net is the input to the neuron.

There are two typical activation functions, commonly used, described by the sigmoidal functions:

the unipolar

$$f(net, \lambda) = \frac{1}{1 + \exp(-\lambda\, net)} \tag{6.1}$$

the bipolar

$$f(net, \lambda) = \frac{2}{1 + \exp(-\lambda\, net)} - 1 \tag{6.2}$$

where $\lambda > 0$ is the gain parameter, $1/\lambda$ being sometimes called the temperature of the system. This parameter describes the slope of the activation functions. These functions are shown in the following Fig. 6.1 and 6.2.

We can observe how a curve changes with respect to the gain parameter λ. Assuming reasonable values of the input, for large value of the gain parameter, the sigmoidal function turns into the Heaviside's function (the step function) while, for small values of the parameter, the sigmoidal function becomes an "almost linear" function.

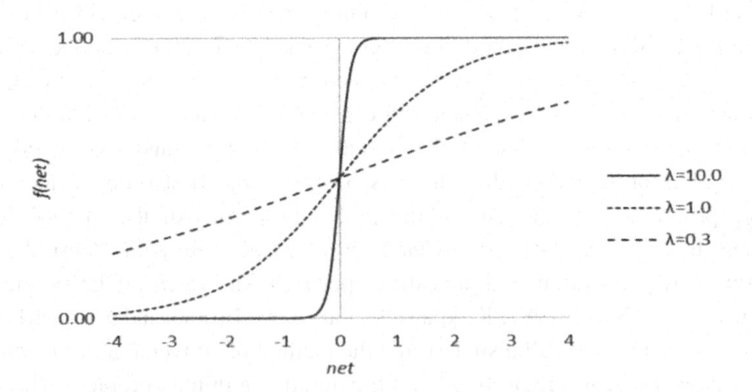

Fig. 6.1 A unipolar sigmoidal function with different gain values

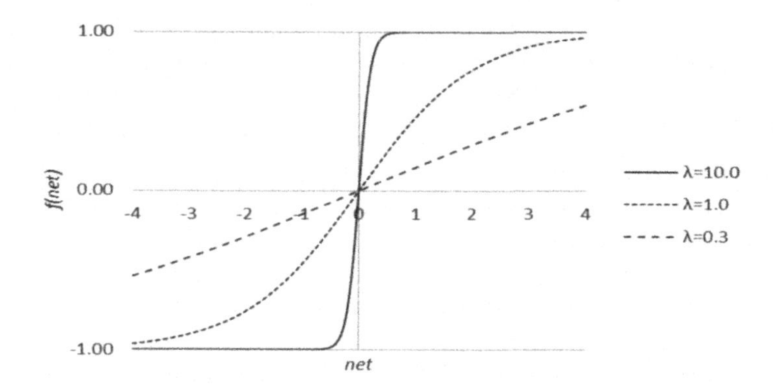

Fig. 6.2 A bipolar sigmoidal function with different gain values

6.3 Continuation Method

In the literature on nonlinear optimisation one can find a little known methodology called *the continuation method* (Avila 1974, Richter and de Carlo 1984), which allows finding *near optimal* solutions. The method is usually presented in terms of finding zeroes of a mapping $G: R^n \rightarrow R$, and it can be generalized for findings fixed points (Ortega and Rheinboldt 1970).

Following the papers of the present author (Krawczak 1999b, 2000b, 2000c, 2000e) we will consider the learning of the neural networks problem as the optimisation problem

$$\min_{W[0,\, L-1]} \left\{ E = \frac{1}{2} \sum_{p=1}^{P} \left\| D_p - X_p(L) \right\|^2 \right\} \tag{6.3}$$

subject to

$$X(l+1) = F(W(l),\, X(l)) = F(l) \quad \text{for} \quad l = 0,1,\ldots,L-1 \tag{6.4}$$

where $F(l)$ is an aggregated function of transition from one layer to another. Let $\tilde{W}[0,\, L-1]$ be a solution of the problem (6.3)-(6.4). Now, let us consider the following homotopy

$$\min_{W[0,\, L-1]} \left\{ E = \frac{1}{2} \sum_{p=1}^{P} \left\| D_p - X_p(L) \right\|^2 \right\} \tag{6.5}$$

subject to

$$X(l+1) = F(W(l),\, X(l),\, \lambda), \quad \text{for} \quad l = 0,1,\ldots,L-1 \tag{6.6}$$

such that for $\lambda = 1$ the problem (6.5)-(6.6) is equivalent to (6.3)-(6.4). It is assumed that the problem (6.5)-(6.6) has some trivial, or easy to compute, solution for $\lambda = \lambda_0$. The value of the parameter λ_0 is treated as the starting point. Additionally, due to the "almost" linear-quadratic form of the considered problem, it is assumed that the solution obtained for $\lambda = \lambda_0$ is the global one. This assumption is based on the properties of the first order differential dynamic methodology described in Chap. 5.

Next, the problem can be imbedded into a family of problems with the parameter λ. Usually we do not consider the gain parameter, which means that we assume that the parameter is equal 1. The basic idea is as follows, if for all $\lambda \in [\lambda_0, 1]$ there exist weights $W(\lambda, [0, L-1])$ such that there are the solutions of (6.5)-(6.6), then the curve $W(\lambda, [0, L-1])$ can be found numerically starting at the point $W(\lambda_0, [0, L-1])$ and ending at $W(1, [0, L-1]) = \tilde{W}(1, [0, L-1])$. In the case of the linear-quadratic optimisation problem, like the neural networks learning, it is obvious that there does exist a solution of (6.5)-(6.6) for any $\lambda \in [\lambda_0, 1]$, where $0 < \lambda_0 \leq 1$. It is assumed that the curve $W(\lambda, [0, L-1])$ is continuous and has first derivatives.

This kind of approach is known in the literature as the continuation method (Avila 1974) or the homotopy method (Richter and de Carlo 1984). Due to the continuous differentiability of the considered sigmoidal activation functions, and the methodology described in the previous chapter, it is obvious that the function $W(\lambda, [0, L-1])$ is continuous and differentiable with respect to the gain parameter $0 < \lambda < +\infty$.

There are two main methods of finding the solution curve $W(\lambda, [0, L-1])$: the first is the discrete method (Avila 1974) and the second – is the Davidenko's method (Davidenko 1953).

Discrete Method

The interval $[\lambda_0, 1]$ should be divided into several segments

$$\lambda_0 < \lambda_1 < \lambda_2 < \cdots \lambda_N = 1$$

and the corresponding partial problems

$$\min_{W[0, L-1]} \left\{ E = \frac{1}{2} \sum_{p=1}^{P} \left\| D_p - X_p(L) \right\|^2 \right\} \tag{6.7}$$

subject to

$$X(l+1) = F(W(l), X(l), \lambda_k), \quad l = 0,1,\ldots,L-1 \tag{6.8}$$

for $k = 0,1,\ldots,N$, should be solved.

Starting with the known solution $W(\lambda_0, [0, L-1])$, obtained by applying the first order differential dynamic programming method, described in Sect. 5.6.1, the new weights $W(\lambda_1, [0, L-1])$ are computed. The procedure is performed sequentially for all $\lambda_0 < \lambda_1 < \lambda_2 < \cdots \lambda_N = 1$ until the point $\lambda_N = 1$ is reached. The main problem is to determine the conditions of existence of the partition $\lambda_0 < \lambda_1 < \lambda_2 < \cdots \lambda_N = 1$ and the iterative process

$$\lambda_{k+1} = I(\lambda_k, W(k, [0, L-1])), \quad k = 0,1,\ldots,N. \tag{6.9}$$

According to Ortega and Rheinboldt (1970), for linear-quadratic optimisation problems the relationship (6.9) exists, and the problem of the partition of the range $[\lambda_0, 1]$ must ensure that the reachable sets of the obtained solution of (6.7)-(6.8)

$$\{W(\lambda_k, [0, L-1]), W(\lambda_{k+1}, [0, L-1])\}$$

and

$$\{W(\lambda_{k+1}, [0, L-1]), W(\lambda_{k+2}, [0, L-1])\}$$

for $k = 0,1,2,\ldots,N-2$ overlap for each pair $\{\lambda_k, \lambda_{k+1}\}$, $k = 0,1,\ldots,N-1$.

Davidenko's Method

The method requires writing the optimisation problem in the following way

$$H(W(\lambda, [0, L-1]), \lambda) = 0. \tag{6.10}$$

By differentiating (6.10) with respect to λ one can get the following Davidenko's differential equation

$$H_W(W(\lambda, [0, L-1], \lambda))\frac{dW(\lambda, [0, L-1], \lambda)}{d\lambda} + H_W(W(\lambda, [0, L-1], \lambda)) = 0 \tag{6.11}$$

with $W(\lambda_0, [0, L-1], \lambda_0)$ as the initial conditions, and by numerical integration from λ_0 to 1.0 - the solution $\tilde{W}[0, L-1] = W(1, [0, L-1])$. The main difficulty is the implicit nature of this equation for such complex problems as the neural network learning.

The proposed application of the continuation method, combined with the first order differential dynamic optimisation approach, gives the new possibility of finding the global optimal value of the performance index of the learning process for the multilayer feedforward neural networks. In some sense the approach is similar to the idea of simulated annealing methodology (Aarts and Laarhoven 1987).

6.4 Heuristic Dynamic Programming

Let us rewrite the here considered first order differential dynamic programming method, described in Chap. 5, and introduce the gain parameter into the neuron models, as follows.

Now, the return function has the form

$$V(X(0), W[0, L-1], \lambda) = \frac{1}{2} \sum_{p=1}^{P} \left\| D_p - X_p(L) \right\|^2. \tag{6.12}$$

The first order expansion of the return function for the whole network $V(X(0), W[0, L-1], \lambda)$ about some nominal weights $\overline{W}[0, L-1]$ and a nominal parameter $\overline{\lambda}$ can be written as

$$\begin{aligned} &V(X'(0), W'[0, L-1], \lambda') = \\ &V(\overline{X}(0), \overline{W}[0, L-1], \overline{\lambda}) + \frac{\partial V(\overline{X}(0), \overline{W}[0, L-1], \overline{\lambda})}{\partial W[0, L-1]} \delta W[0, L-1] \\ &+ \frac{\partial V(\overline{X}(0), \overline{W}[0, L-1], \overline{\lambda})}{\partial \lambda[0, L-1]} \delta\lambda \,. \end{aligned} \tag{6.13}$$

The variations in the weight variables $\delta W = W' - \overline{W}$ and the parameter $\delta\lambda = \lambda' - \overline{\lambda}$ must be small enough to ensure the validity of the expansion. Choosing δW and $\delta\lambda$ as

$$\delta W[0, L-1] = \eta \left[\frac{\partial V(\overline{X}(0), \overline{W}[0, L-1], \overline{\lambda})}{\partial W[0, L-1]} \right]^T \tag{6.14}$$

and

$$\delta\lambda = \eta \left[\frac{\partial V(\overline{X}(0), \overline{W}[0, L-1], \overline{\lambda})}{\partial \lambda} \right]^T \tag{6.15}$$

where η is the learning parameter, we obtain that the return function

$$V(X'(0), W'[0, L-1], \overline{\lambda})$$

will be smaller than

$$V(\overline{X}(0), \overline{W}[0, L-1], \overline{\lambda}).$$

Instead of consideration of the return function for the whole network $V(X(0, W[0, L-1], \lambda))$, the backward form can be calculated

$$V(X(l), W[l, L-1], \lambda) = V(X(l+1), W[l+1, L-1], \lambda) \tag{6.16}$$

for l, $l = 1,2,..., L-1$, with the condition for the last layer

$$V(X(L), W[L-1, L-1], \lambda) = \frac{1}{2}\sum_{p=1}^{P}\left\|D_p - X_p(L)\right\|^2. \tag{6.17}$$

The partial derivative of the return function (6.16) takes the form

$$V_W(l) = F_W^T(l)\, V_X(l+1, \lambda)\, F_W(l). \tag{6.18}$$

In order to obtain (6.18) the following expression must be calculated $V_X(l, \lambda)$, which can be calculated using the following sequential relations

$$V_X(L, \lambda) = -\sum_{p=1}^{P}\left\|D_p - X_p(L)\right\| \tag{6.19}$$

$$V_X(l, \lambda) = F_X(l)\, V_X(l+1, \lambda). \tag{6.20}$$

The derivative of the return function $V(X(0, W[0, L-1], \lambda))$ with respect to λ is obtained in a similar manner, as

$$V_\lambda(L, \lambda) = -\sum_{p=1}^{P}\left\|D_p - X_p(L)\right\| \tag{6.21}$$

$$V_\lambda(l, \lambda) = V_\lambda(l+1, \lambda) + F_\lambda(l)\, V_X(l+1, \lambda) \tag{6.22}$$

where $V_X(l+1, \lambda)$ is given by (6.19) and (6.20).

Using Equ. 6.19 - 6.21, the gradients of the return functions, which are required for Equ. 6.18 and 6.22, can be computed from the sequence of the equations solved from the last layer back to the inputs.

The parameter λ can be found at the input layer by minimizing the optimal return function $V\left(X'(0), W'[0, L-1], \bar{\lambda}\right)$ with respect to λ.

The heuristic differential dynamic programming algorithm for the learning of multilayer neural networks can be formulated in the following steps:

1. Initialise weights: $\overline{W}(l)$, $l = 0,1,2,..., L-1$
   $\bar{\lambda}$ as a small value, e.g. 0.1
2. Set: $E = 0$
3. Set: $p = 0$ (p denotes the pattern's index)
4. Submit a pattern: $(X_p(0), D_p)$, $p = p+1$

5. Compute the layers' outputs: $X(l) = \overline{X}(l), \quad l = 1,2,\ldots, L$ from the system Equ. 6.8

6. Compute the performance index of learning:

$$E = E + \frac{1}{2} \left\| D_p - \overline{X}_p(L) \right\|^2 \tag{6.23}$$

7. Compute the partial derivatives:

$$V_X(l) = V_X\left(\overline{X}(l), \overline{W}[l, L-1], \overline{\lambda}\right) \text{ for } l = L, L-1,\ldots, 0 \tag{6.24}$$

from

$$V_X(l) = F_X(l) V_X(l+1) \tag{6.25}$$

$$V_X(L) = -\left\| D_p - X_p(L) \right\|$$
$$V_\lambda(l) = V_\lambda(l+1) + F_\lambda(l) V_X(l+1) \tag{6.26}$$

$$V_\lambda(L) = -\sum_{p=1}^{P} \left\| D_p - X_p(L) \right\|$$

8. Compute the gradient of the return function with respect to the weights for each level:

$$\frac{\partial V\left(\overline{X}(l), \overline{W}[l, L-1], \overline{\lambda}\right)}{\partial \overline{W}(l)} = \left[\frac{\partial F\left(\overline{X}(l), \overline{W}(l)\right)}{\partial \overline{W}(l)}\right]^T \frac{\partial V\left(\overline{X}(l+1), \overline{W}[l+1, L-1], \overline{\lambda}\right)}{\partial \overline{X}(l+1)} \tag{6.27}$$

for $l = L-1,\ldots,0$

9. Choose the learning parameter $\eta > 0$ and compute the new weight values

$$W'(l) = \overline{W}(l) - \eta \left[\frac{\partial V\left(\overline{X}(l), \overline{W}[l, L-1], \overline{\lambda}\right)}{\partial \overline{W}(l)}\right]^T \tag{6.28}$$

10. If $p = P$ then go to Step 3 else go to Step 11

11. If $l=0$ then compute the new value of the gain parameter, for $\eta_1 > 0$

$$\lambda' = \bar{\lambda} - \eta_1 \left[\frac{\partial V\left(\bar{X}(0), \bar{W}[0, L-1], \bar{\lambda}\right)}{\partial \lambda} \right]^T \tag{6.29}$$

12. Set:

$\bar{W}[0, L-1] = W'[0, L-1]$

$\bar{X}[1, L] = X'[1, L]$

$\bar{\lambda} = \lambda'$

13. If $E < E_{MAX}$ then go to Step 14 else go to Step 3

14. STOP.

The first order differential dynamic programming algorithm can ensure "almost global" optimisation of the learning process for the neural networks with the linearized neurons. By optimisation of the gain parameter the methodology allows for finding of the "almost global" optimum of the learning index performance. We will call this method the heuristic dynamic programming for the multilayer neural networks learning.

6.5 Learning as a Multiobjective Problem

The idea of using multiobjective optimisation to solve some classes of nonlinear programming problems was first proposed by Geoffrion (1967). This idea was extended by Li and Haimes (1990, 1991) for the dynamic systems. In the case of the neural networks learning, the idea was further elaborated by Krawczak (1995b, 1997, 1998).

6.5.1 Embedding into Multiobjective Optimisation Problems

It is easy to notice that the performance index of learning, E, can be treated as a simple composite function of multiple performance indices E_p, corresponding to the training pairs $\{D_p, X_p(0)\}$, $p = 1,2,\ldots,P$, as follows

$$\min\left\{ E = \Phi(E_1, E_2, \ldots, E_P) = \sum_{p=1}^{P} \left\| D_p - X_p(L) \right\|^2 \right\} \tag{6.30}$$

subject to

$$X_p(l+1) = F\left(X_p(l), W(l)\right), \; l = 0,1,\ldots,L-1. \tag{6.31}$$

The minimization is done with respect to weights $W(l)$, $l = 0,1,\ldots,L-1$. In the discussed case, the overall performance index E is a strictly increasing function of E_p, for $p = 1,\ldots,P$, i.e.

$$\frac{\partial E}{\partial E_p} = 1 .$$

Conforming to the notation used, the multiobjective multistage optimisation problem can be formulated in the following way

$$\min \begin{bmatrix} E_1 \\ \vdots \\ E_P \end{bmatrix} = \min \begin{bmatrix} \|D_1 - X_1(L)\|^2 \\ \vdots \\ \|D_p - X_p(L)\|^2 \end{bmatrix} \tag{6.32}$$

subject to

$$X_p(l+1) = F(X_p(l), W(l)), \quad l = 0,1,\ldots,L-1. \tag{6.33}$$

In general, solution to the multiobjective problem (6.32)-(6.33) is not unique. A solution $\hat{W}$ of this problem is said to be noninferior if there does not exist another feasible W such that

$$E_p(W) \le E_p(\hat{W}) \tag{6.34}$$

for all $p = 1,2,\ldots,P$, with strict inequality for at least one p.

For the optimisation problem (6.32)-(6.33) the following theorems can be proved (Li and Haimes 1990). Here, the respective analysis is developed for the neural networks learning problem.

Theorem 6.1. *The optimal solution of problem (6.30)-(6.31) is attained by a noninferior solution of the multiobjective optimisation problem given through (6.32)-(6.33).*

The most common approaches to generation of the set of noninferior solutions are the ε-constraint method and the weighting method (Li and Haimes 1990).

The noninferior solutions to the problem (6.32)-(6.33) can be generated by solving the following ε-constraint method form

$$\min E_1(W) \tag{6.35}$$

subject to

$$E_p(W) \le \varepsilon_p, \; p = 2,3,\ldots,P \tag{6.36}$$

and

$$X_p(l+1) = F(X_p(l), W(l)), \quad l = 0,1,\ldots,L-1 \tag{6.37}$$

where $X_p(0)$ is given.

Theorem 6.2. *Assume that the set of noninferior solutions of problem (6.32-(6.33) can be parameterised by* μ_{1p}, $p = 2,3,\ldots,P$, *which are the optimal Kuhn-Tucker multipliers associated with the p-th constraint in Equ. (6.36. Thus, the optimal solution of the dynamic optimisation problem (6.30)-(6.31) is then reached by the noninferior solution that satisfies the following equalities*

$$\frac{\partial E}{\partial E_p} - \mu_{1p} \frac{\partial E}{\partial E_1} = 0, \quad p = 2,3,\ldots,P. \tag{6.38}$$

For the case considered in this section, the objective function E is of an additive form, and so the solution to Equ. (6.30)-(6.31) is attained by the noninferior solution with all μ_{1p} equal to one.

If problem (6.32)-(6.33) is convex, then the noninferior solutions of problem (6.30)-(6.31) can be obtained by solving the following weighting form

$$\min \sum_{p=1}^{P} v_p E_p(W) \tag{6.39}$$

subject to

$$X_p(l+1) = F(X_p(l), W(l)), \quad l = 0,1,\ldots,L-1 \tag{6.40}$$

where $X_p(0)$ is given, and

$$v_p \geq 0,\ p = 1,2,\ldots,P;\ \sum_{p=1}^{P} v_p = 1 .$$

Theorem 6.3. *If the set of noninferior solutions of the problem (6.32)-(6.33) can be parameterised by the overall weighting vector* v, *the optimal solution of the nonseparable dynamic optimisation problem given in (6.30)-(6.31) is reached under certain conditions by the noninferior solution that satisfies the following equations*

$$\frac{\partial E / \partial E_1}{v_1} = \frac{\partial E / \partial E_2}{v_2} = \cdots = \frac{\partial E / \partial E_P}{v_P}. \tag{6.41}$$

For the discussed case, due to the additive form of the performance index, the optimal solution of problem (6.30)-(6.31) is attained by the noninferior solution with all v_p equal to $1/P$.

The aim of consideration of the optimisation problem (6.30)-(6.31) as an embedded problem, meaning the embedding of this optimisation problem in a family of parameterised optimisation problems, is to obtain a new algorithm for the multilayer neural networks learning.

Let us consider the following weighted minimax formulation for problem (6.30)-(6.31):

$$\min \max\{v_1 E_1, v_2 E_2, \ldots, v_P E_P\} \tag{6.42}$$

subject to

$$X_p(l+1) = F(X_p(l), W(l)), \tag{6.43}$$

where $X_p(0)$ is given, the weighting coefficient v_1 is always set to one and the weighting coefficients v_p, $p = 2, \ldots, P$ are nonnegative. In (6.42) the maximization is performed among P weighted systems indicated by the performance indices (corresponding to any training pair) while minimization is carried out over the weights searching W.

6.5.2 Iterative Minimax Solution

Several theorems strictly related to the conversion of the optimisation problem (6.30)-(6.31) into another optimisation problem, (6.42)-(6.43), can be proven. If we denote by W^* the set of solutions of the problem (6.30)-(6.31) and by W_v^* the union of sets of solutions of the weighted minimax problem (6.42)-(6.43), then it is possible to prove the following

Theorem 6.4. *The intersection of* W^* *and* W_v^* *is nonempty, i.e.*

$$W^* \cap W_v^* = \varnothing.$$

The consequence of this theorem is exceptionally clear, namely if the optimisation problem (6.30)-(6.31) has a unique solution, then this optimal solution can be generated by the weighted minimax solution of (6.42)-(6.43). In the case of existence of multiple solutions to (6.30)-(6.31) then at least a nonempty subset of W_v^* can be generated by the weighted minimax problem (6.42)-(6.43).

It is possible to show that the minimax optimisation problem (6.42)-(6.43) is equivalent to the following form

$$\min \varphi(y) \tag{6.44}$$

subject to

$$v_p E_p(W) \leq y, \; p = 1, 2, \ldots, P$$

$$X_p(l+1) = F(X_p(l), W(l)), \quad l = 0, 1, \ldots, L-1 \tag{6.45}$$

$$X_p(0) \text{ is given}$$

where y is an auxiliary variable, while $\varphi(y)$ is any increasing differentiable function.

By minimizing a strictly increasing function of y we will minimize the maximum value among $v_1E_1, v_2E_2, \cdots, v_PE_P$.

For positive Kuhn-Tucker multipliers occurring in the Lagrangian of the problem (6.44)-(6.45), this problem can be rewritten as

$$\min \varphi(y) \tag{6.46}$$

subject to

$$\begin{aligned} &-\frac{y}{E_p(W)} \le -v_p, \; p = 1,2,\ldots,P \\ &X_p(l+1) = F\left(X_p(l), W(l)\right), \; l = 0,1,2,\ldots,L-1 \end{aligned} \tag{6.47}$$

with $X_p(0)$ given.

The methodology proposed could be used for deriving a new paradigm of learning, which we will call *updating of learning*.

6.6 Updating of Learning

The methodology described can be used for updating in the learning process of the neural networks. After finishing the learning of a neural network using P examples, when there is a new example $P+1$, we can treat the updating of learning as the two-objective optimisation problem with two objective functions E_1 - obtained for P examples, and E_2 - related to the new example $P+1$.

The problem of updating learning can be stated as follows. The performance index of learning E of the form

$$E = \Phi(E_1, E_2) = \sum_{p=1}^{P} \left\| D_p - X_p(L) \right\|^2 + \left\| D_{P+1} - X_{P+1}(L) \right\|^2 \tag{6.48}$$

is minimized subject to:

for $\{D_p, X_p(0)\}$, $p = 1,2,\ldots,P$:

$$X_p(l+1) = F\left(X_p(l), W(l)\right), \quad l = 0,1,\ldots,L-1 \tag{6.49}$$

and for $\{D_{P+1}, X_{P+1}(0)\}$:

$$X_{P+1}(l+1) = F\left(X_{P+1}(l), W(l)\right), \quad l = 0,1,\ldots,L-1. \tag{6.50}$$

The overall performance index E is a strictly increasing function of E_1 and E_2, i.e.

$$\frac{\partial E}{\partial E_1}=\frac{\partial E}{\partial E_2}=1.$$

The two-objective multistage optimisation problem can be formulated as follows

$$\min\begin{bmatrix}E_1\\E_2\end{bmatrix}=\min\begin{bmatrix}\sum_{p=1}^{P}\left\|D_p-X_p(L)\right\|^2\\ \left\|D_{P+1}-X_{P+1}(L)\right\|^2\end{bmatrix} \tag{6.51}$$

subject to

for $\{D_p, X_p(0)\}$, $p=1,2,\ldots,P$:

$$X_p(l+1)=F(X_p(l),W(l)),\quad l=0,1,\ldots,L-1 \tag{6.52}$$

and for $(D_{P+1}, X_{P+1}(0))$:

$$X_{P+1}(l+1)=F(X_{P+1}(l),W(l)),\quad l=0,1,\ldots,L-1. \tag{6.53}$$

The derivation of the procedure for the updating of learning of the neural networks can be continued following Sect. 6.4.

Even if the procedure described above seems to be complex, for extremaly large neural networks this way of weight updating may be worth of interest in order to make the learning time shorter.

6.7 Description of Heuristic Dynamic Programming Algorithm

In Sect. 6.2 we have introduced the gain parameter λ into the artificial neuron models. The changes of this parameter allow for changing of the "nonlinearity" of the neuron activation functions. For small values of the parameter, the sigmoidal activation function becomes an "almost linear" function. This property has been used to construct the heuristic dynamic programming algorithm based on the first order differential dynamic programming algorithm. In this section, we will extend the generalized net description of the first order differential dynamic programming algorithm by including one extra transition responsible for changes of the gain parameter. This section is based on the papers by Krawczak and Aladjov (2002), Krawczak (2002b, 2002c, 2003a, 2003b).

The full representation will be introduced with a short repetition of the common parts of the description described in Sect. 5.5.

The α-type tokens enter the net through the place $\ddot{X}_1$ and have the initial characteristics

$$y(\alpha_{i(l)})=\langle NN1,\, l,\, i(l),\, f_{i(l)}\rangle \text{ for } i(l)=1,2,\ldots,N(l),\ l=0,1,\ldots,L \tag{6.54}$$

where

$NN1$
the neural network identifier,

$i(l)$
the number of the token associated with the l -th layer,

l
the present layer number,

$f_{1(l)}(\cdot)$
an activation function of the i -th neuron associated with the l -th layer of the neural network, where $f_{i(0)}(\cdot)=1.0$, for $i(0)=1,2,\ldots,N(0)$.

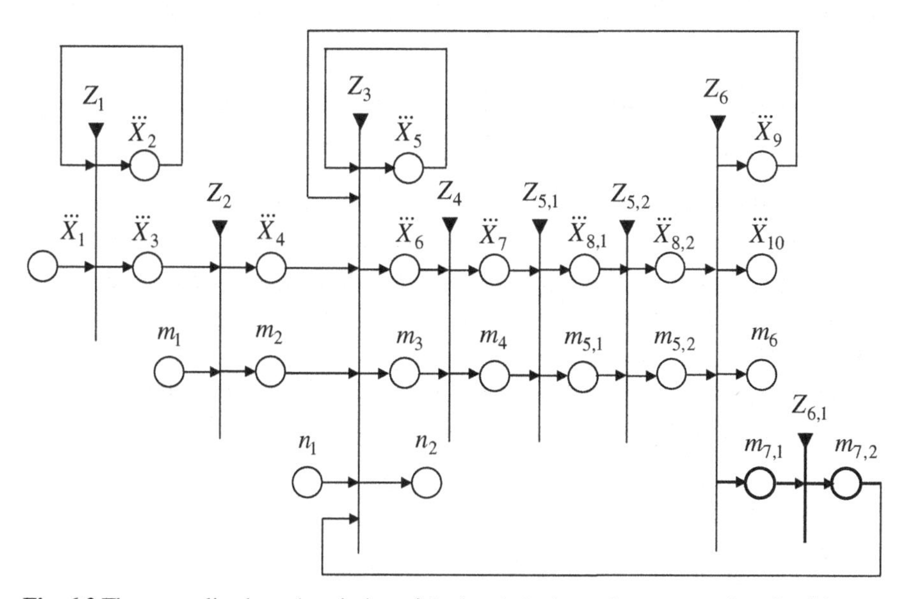

Fig. 6.3 The generalized net description of the heuristic dynamic programming algorithm

The generalized net representation of the heuristic dynamic programming algorithm contains eight transitions, see Fig. 6.3, which will be described below.

The tokens $\alpha_{i(l)}$, $i(l)=1,2,\ldots,N(l)$, $l=0,1,\ldots,L$, are transferred from the place $\dddot{X}_1$ to the place $\dddot{X}_2$ as well as $\dddot{X}_3$. The transition Z_1 has the form

$$Z_1=\langle\{\dddot{X}_1,\dddot{X}_2\},\{\dddot{X}_2,\dddot{X}_3\},\quad \begin{array}{c|cc} & \dddot{X}_2 & \dddot{X}_3 \\ \hline \dddot{X}_1 & V_{1,2} & V_{1,3} \\ \dddot{X}_2 & V_{2,2} & V_{2,3} \end{array},\vee(\dddot{X}_1,\dddot{X}_2)\rangle, \tag{6.55}$$

where

$V_{1,2} = \neg V_{1,3} =$ "if there is only one token $\alpha_{i(l)}$ in the place $\dddot{X}_1$", i.e.

$$\forall\left(\alpha_{j(l)} \in K_{\dddot{X}_1}\right)\left(pr_2 Y_{\alpha_{i(l)}} \neq pr_2 Y_{\alpha_{j(l)}}, j(l) \neq i(l)\right) \tag{6.56}$$

(where $K_{\dddot{X}_1}$ is a set of all tokens entering the net from the place $\dddot{X}_1$, $i(l), j(l) = 1, 2, \ldots, N(l)$)

$V_{2,2} =$ "if there is more than one token $\alpha_{i(l)}$ and $\alpha_{j(l)}$ associated with the l-th layer", i.e.,

$$\exists\left(\alpha_{i(l)} \in K_{\dddot{X}_1} \,\&\, \alpha_{j(l)} \in K_{\dddot{X}_2}\right)\left(pr_2 Y_{\alpha_{i(l)}} = pr_2 Y_{\alpha_{j(l)}}\right) \tag{6.57}$$

$V_{2,3} =$ "if all tokens $\alpha_{i(l)}$, $i(l) = 1, 2, \ldots, N(l)$, have been combined into one token" i.e.,

$$\begin{aligned} &\neg\exists\left(\alpha_{i(l)} \in K_{\dddot{X}_1} \,\&\, \alpha_{k(l)} \in K_{\dddot{X}_2}\right)\left(pr_2 Y_{\alpha_{i(l)}} = pr_2 Y_{\alpha_{k(l)}}\right) \& \\ &\neg\exists\left(\alpha_{i(l)} \in K_{\dddot{X}_1} \,\&\, \alpha_{j(l)} \in K_{\dddot{X}_1}\right)\left(pr_2 Y_{\alpha_{i(l)}} = pr_2 Y_{\alpha_{j(l)}}, i \neq j\right). \end{aligned} \tag{6.58}$$

In general, the gain parameter λ was incorporated in the models (6.1) and (6.2) of the considered neuron activation functions. Here, the transition Z_1 represents only the conceptual model of the neural network architecture, and the parameters are defined describing the number of layers L, and the numbers of neurons $N(l)$ within each layer $l = 0, 1, \ldots, L$.

The new token $\alpha_{(l)}$ representing the l-th layer is transferred from the place $\dddot{X}_2$ to the place $\dddot{X}_3$, and obtains the characteristic

$$y\left(\alpha_{(l)}\right) = \langle NN1, l, [1, N(l)], F_{(l)} \rangle \quad \text{for } l = 0, 1, 2, \ldots, L, \tag{6.59}$$

where

$NN1$
the neural network identifier,

l
the layer number,

$[1, N(l)]$
denotes $N(l)$ tokens (neurons) arranged in a sequence, starting form the first and ending at $N(l)$, associated with the l-th layer,

$F_{(0)} = [1, 1, \ldots, 1]^T$, $F_{(l)} = \left[f_{1(l)}(\cdot), f_{2(l)}(\cdot), \ldots, f_{N(l)}(\cdot)\right]^T$
is a vector of the activation functions of the neurons associated with the l-th layer of the neural network.

The transition Z_2 has the following form

$$Z_2 = \langle \{\dddot{X}_3, m_1\}, \{\dddot{X}_4, m_2\}, \begin{array}{c|cc} & \dddot{X}_4 & m_2 \\ \hline \dddot{X}_3 & true & false \\ m_1 & false & true \end{array}, \wedge(\dddot{X}_3, m_1) \rangle \tag{6.60}$$

and the information of the performance index is associated with the β-type token, the token enters the input place m_1 with the initial characteristic

$$y(\beta) = \langle NN1, E, E_{\max}, \lambda \rangle \tag{6.61}$$

where

$NN1$
the neural network identifier,

E
the performance index of the neural network learning,

$E_{\max}$
the threshold value of the performance index, which must be reached.

The token $\alpha_{(l)}$, $l = 0,1,2,...,L$, obtains the following characteristic in the place $\dddot{X}_4$

$$y(\alpha_{(l)}) = \langle NN1, l, [1, N(l)], F_{(l)}, W_{(l)}, \lambda \rangle, \quad \text{for } l = 0,1,2,...,L, \tag{6.62}$$

where

$NN1$
is the neural network identifier,

l
the layer number,

$[1, N(l)]$
denotes $N(l)$ tokens (neurons) arranged in a sequence, starting form the first and ending at $N(l)$, associated with the l-th layer,

$F_{(0)} = [1,1,...,1]^T$, $F_{(l)} = [f_{1(l)}(\cdot), f_{2(l)}(\cdot),..., f_{N(l)}(\cdot)]^T$
is a vector of the activation functions of the neurons associated with the l-th layer,

$W_{(l)}$
denotes the aggregated initial weights connecting the neurons of the l-th layer with the $(l-1)$-st layer neurons,

λ
denotes the initial value of the gain parameter (e.g. $\bar{\lambda} = 0.1$).

In place m_2 the β token obtains the characteristic

$$y(\beta) = \langle NN1, 0, E_{\max}, \bar{\lambda} \rangle. \tag{6.63}$$

In the transition Z_3 the tokens γ_p, $p = 1,2\ldots,P$, enter the place n_1 with the initial characteristic $y(\gamma_p) = \langle X_p(0), D_p, p \rangle$, where $X_p(0) = [x_{p1}, x_{p2}, \ldots, x_{pN(0)}]^T$ the inputs vector of the neural network, $D_p = [d_{p1}, d_{p2}, \ldots, d_{pN(0)}]^T$ the vector of desired network outputs, and for the input $X_p(0)$ the outputs of all layers are calculated.

The transition Z_3 associated with the signal propagation process in the neural network has the form

$$Z_3 = \langle \{\dddot{X}_4, \dddot{X}_5, \dddot{X}_9, m_2, m_7, n_1\}, \{\dddot{X}_5, \dddot{X}_6, m_3, n_2\},$$

	$\dddot{X}_5$	$\dddot{X}_6$	m_3	n_2
$\dddot{X}_4$	$V_{4,5}$	$V_{4,6}$	*false*	*false*
$\dddot{X}_5$	$V_{5,5}$	$V_{5,6}$	*false*	*false*
$\dddot{X}_9$	$V_{9,5}$	$V_{9,6}$	*false*	*false*
m_2	*false*	*false*	*true*	*false*
$m_{7,2}$	*false*	*false*	*true*	*false*
n_1	*false*	*false*	*false*	*true*

$$\wedge(\vee(\dddot{X}_4, \dddot{X}_5, \dddot{X}_9), (m_2, m_{7,2}), n_1)\rangle \tag{6.64}$$

where

$V_{4,5} = V_{5,5} = V_{9,5} =$ "previous layer does not have defined outputs",

$V_{4,6} = V_{5,6} = V_{9,6} = \neg V_{4,5}$,

$V_{1,2} =$ "all layers' outputs have assigned values for the current pattern".

The tokens $\alpha_{(l)}$, $l = 0,1,2,\ldots,L$, in the place $\dddot{X}_5$ obtain the characteristics

$$y(\alpha_{(l)}) = \langle NN1, l, [1, N(l)], F_{(l)}, W_{(l)}, X_{(l)}, \lambda \rangle \tag{6.65}$$

where $X_{(l)}$, $l = 1,2,\ldots,L$, is the vector of neuron outputs related to the l-th layer due to $W_{(l)}$, $l = 1,2,\ldots,L$. In the place $\dddot{X}_6$ the tokens get the following characteristics

$$y(\alpha_{(l)}) = \langle NN1,\ l,\ [1, N(l)],\ F_{(l)}, \overline{W}_{(l)}, \overline{X}_{p(l)}, \overline{\lambda} \rangle \tag{6.66}$$

for the nominal values of the weights $\overline{W}_{(l)}$, states $\overline{X}_{p(l)}$ and the gain parameter $\overline{\lambda}$. The token β in the place m_3 has the characteristic $y(\beta) = \langle NN1, 0, E_{\max}, \overline{\lambda} \rangle$, and in the place n_2 the token γ has the characteristic $y(\gamma_p) = \langle X_p(0),\ D_p,\ p \rangle$.

The transition Z_4 is responsible for the estimation and weights adjustment, and has the form

$$Z_4 = \langle \{\dddot{X}_6,\ m_3\}, \{\dddot{X}_7,\ m_4\}, \quad \begin{array}{c|cc} & \dddot{X}_7 & m_4 \\ \hline \dddot{X}_6 & true & false \\ m_3 & false & true \end{array} \quad, \wedge(\dddot{X}_6, m_3) \rangle . \tag{6.67}$$

In the place m_4 the token β has the characteristic, which determines the performance index $y(\beta) = \langle NN1, E', E_{\max}, \overline{\lambda} \rangle$, where

$$E' = E + \frac{1}{2} \sum_{j(L)=1}^{N(L)} (d_{p,j(L)} - x_{p,j(L)})^2 = pr_2 \langle NN1, E, E_{\max} \rangle^{\beta} + \frac{1}{2} \sum_{j(L)=1}^{N(L)} (d_{p,j(L)} - x_{p,j(L)})^2 .$$

Similarly as in the previous section, the transition Z_5 is split in two transitions $Z_{5,1}$ and $Z_{5,2}$. In this case the transition $Z_{5,1}$ models the evaluation of the partial derivatives of the return functions $V(X(l), W[l, L-1], \lambda)$ with respect to $X(l)$ and additionally to λ

$$Z_{5,1} = \langle \{\dddot{X}_7,\ m_4\}, \{\dddot{X}_{8,1},\ m_{5,1}\}, \quad \begin{array}{c|cc} & \dddot{X}_{8,1} & m_{5,1} \\ \hline \dddot{X}_7 & true & false \\ m_4 & false & true \end{array} \quad, \wedge(\dddot{X}_7, m_4) \rangle . \tag{6.68}$$

For the nominal values $\overline{X}(l)$, $\overline{W}[l, L-1]$ for $l = L, L-1, \ldots, 0$, and $\overline{\lambda}$ the derivatives have the form

$$V_X(L) = -\sum_{p=1}^{P} \| D_p - X_p(L) \|$$

$$\overline{V}_X(l) = F_X(l) \overline{V}_X(l+1) \quad \text{for } l = L-1, \ldots, 0 \tag{6.69}$$

$$V_\lambda(L) = -\sum_{p=1}^{P} \| D_p - X_p(L) \|$$

$$\overline{V}_\lambda(l) = \overline{V}_\lambda(l+1) + F_\lambda(l) \overline{V}_X(l+1) \quad \text{for } l = L-1, \ldots, 0. \tag{6.70}$$

The α -type tokens obtain, in the place $\dddot{X}_{8,1}$, the following characteristics

$$y(\alpha_{(l)}) = \langle NN1, l, [1, N(l)], F_{(l)}, \overline{W}_{(l)}, \overline{X}_{p(l)}, \overline{V}_X(l), \overline{V}_\lambda(l), \overline{\lambda} \rangle \tag{6.71}$$

and the characteristic of the β token is not changed in the place $m_{5,1}$.

The transition $Z_{5,2}$ is devoted to the partial derivatives of the return functions $V(X(l), W[l, L-1], \lambda)$ with respect to $W[l, L-1]$, and has the form

$$Z_{5,2} = \langle \{\dddot{X}_{8,1}, m_{5,1}\}, \{\dddot{X}_{8,2}, m_{5,2}\}, \begin{array}{c|cc} & \dddot{X}_{8,2} & m_{5,2} \\ \hline \dddot{X}_{8,1} & true & false \\ m_{5,1} & false & true \end{array}, \wedge(\dddot{X}_{8,1}, m_{5,1}) \rangle \tag{6.72}$$

while the α tokens obtain the following new characteristics, in the place $\dddot{X}_{8,2}$

$$y(\alpha_{(l)}) = \langle NN1, l, [1, N(l)], F_{(l)}, \overline{W}_{(l)}, \overline{X}_{p(l)}, \overline{V}_W(l), \overline{\lambda} \rangle \tag{6.73}$$

where $\overline{V}_W(l) = F_W^T(l)\overline{V}_X(l+1, \lambda)\, F_W(l)$ for $l = L-1, \ldots, 0$, while the token β has not changed the characteristic.

The transition Z_6 describing the weight adjustments has the following form

$$Z_6 = \langle \{\dddot{X}_{8,2}, m_{5,2}\}, \{\dddot{X}_9, \dddot{X}_{10}, m_6, m_{7,1}\}, \begin{array}{c|cccc} & \dddot{X}_9 & \dddot{X}_{10} & m_6 & m_{7,1} \\ \hline \dddot{X}_{8,2} & V_{8,9} & V_{8,10} & false & false \\ m_{5,2} & false & false & V_{5,6} & V_{5,7} \end{array}, \wedge(\dddot{X}_{8,2}, m_{5,2}) \rangle \tag{6.74}$$

where

$V_{8,9} =$ "there are still unused patterns",

$V_{8,10} = \neg V_{8,9}$,

$V_{5,6} =$ "if the performance index is below the given threshold E_{max}",

$V_{5,7} = \neg V_{5,6}$.

The α -type tokens obtain the new characteristics in the place $\dddot{X}_9$

$$y(\alpha_{(l)}) = \langle NN1, l, [1, N(l)], F_{(l)}, W'_{(l)}, \overline{\lambda} \rangle \tag{6.75}$$

with updated weight connections

$$W'_{(l)} = \left[w'_{1(l)}, w'_{2(l)}, \ldots, w'_{N(l)}\right]^T$$

where $w'_{i(l)} = \left[w'_{i(l-1)1(l)}, w'_{i(l-1)2(l)}, \ldots, w'_{i(l-1)N(l)}\right]^T$,
for $i(l-1) = 1, 2, \ldots, N(L-1)$, $j(l) = 1, 2, \ldots, N(L)$.

The new values of the weights are calculated in the following way (for the learning parameter $\eta > 0$)

$$W'(l) = \overline{W}(l) - \eta \left[\frac{\partial V\left(\overline{X}(l), \overline{W}[l, L-1], \overline{\lambda}\right)}{\partial \overline{W}(l)}\right]^T \quad \text{for } l = L-1, \ldots, 0. \tag{6.76}$$

and replace $\overline{W}[0, L-1] = W'[0, L-1]$, $\overline{X}[1, L] = X'[1, L]$.

If the stop condition is satisfied, then the optimal values of the weights are defined as $W^*_{(l)} = pr_5 \langle NN1, l, [1, N(l)], F_{(l)}, W'_{(l)}, \overline{\lambda} \rangle$, where the characteristics of the α-type tokens in the place $\dddot{X}_{10}$ are described by $y(\alpha_{(l)}) = \langle NN1, l, [1, N(l)], F_{(l)}, W'_{(l)}, \overline{\lambda} \rangle$. The final value of the performance index is equal $E^* = pr_2 \langle NN1, E', E_{\max}, \overline{\lambda} \rangle$, where the β token characteristic in the place m_6 is described by $y(\beta) = \langle NN1, E', E_{\max}, \overline{\lambda} \rangle$.

In the place $m_{7,1}$ the β token obtains the characteristic

$$y(\beta) = \langle NN1, E, E_{\max}, \overline{\lambda} \rangle \tag{6.77}$$

which is not final.

The transition $Z_{6,1}$ describing the gain parameter adjustments has the following form

$$Z_{6,1} = \langle \{m_{7,1}\}, \{m_{7,2}\}, \begin{array}{c|c} & m_{7,2} \\ \hline m_{7,1} & true \end{array}, \wedge(m_{7,1}) \rangle. \tag{6.78}$$

The β token obtains the new characteristics in the place $m_{7,2}$

$$y(\beta) = \langle NN1, E, E_{\max}, \lambda' \rangle \tag{6.79}$$

where for a small positive value η_1 the new value of the gain parameter is calculated in the following way

$$\lambda' = \overline{\lambda} - \eta_1 \left[\frac{\partial V\left(\overline{X}(0), \overline{W}[0, L-1], \overline{\lambda}\right)}{\partial \lambda}\right]^T \tag{6.80}$$

and is replaced $\overline{\lambda} = \lambda'$.

The here constructed generalized net representation show the possibilities for the generalized nets to describe the way of functioning and the results of work of real processes. Also, they can be used not only for representing the parallel functioning of homogeneous objects (e.g., as we see above - neural networks), but also for representing the parallel functioning of non-homogeneous objects.

Chapter 7
Adjoint Neural Networks

7.1 Neural Network as a Flow Graph

In the very rich artificial neural networks literature little attention has been given to consideration of neural networks from the point of graph theory. Examination of neural networks *as flow graphs* gives very interesting and new properties of the neural networks learning process. The approach is based on the Tellegen's theorem (Tellegen 1952, Chua and Lin 1975) used in the electric circuits when Kirchhoff's laws may not be valid. Here we use terminology adopted from the optimisation or optimal control theory (Bryson and Ho 1969), and such neural networks in which signals flow in opposite direction are called *the adjoint neural networks.*

In this section we present a flow graph methodology for representing neural networks. Since the pioneering work of McCulloch and Pitts (1943) a model of an artificial neuron is a very simple processing unit described depicted in Fig. 1.6, here denoted as Fig. 7.1, which has a number of inputs, $x_{i(l-1)}$, $i(l-1) = 1,2,\dots, N(l-1)$.

The sum of the weighted inputs and the bias (included in the inputs) forms at the summation point $\oplus$ the following signal

$$net_{j(l)} = \sum_{i(l-1)=1}^{N} w_{i(l-1)j(l)}\, x_{i(l-1)} = \sum_{i(l-1)=1}^{N} y_{i(l-1)j(l)} \tag{7.1}$$

as an argument of an activation function $f_{j(l)}\left(net_{j(l)}\right)$.

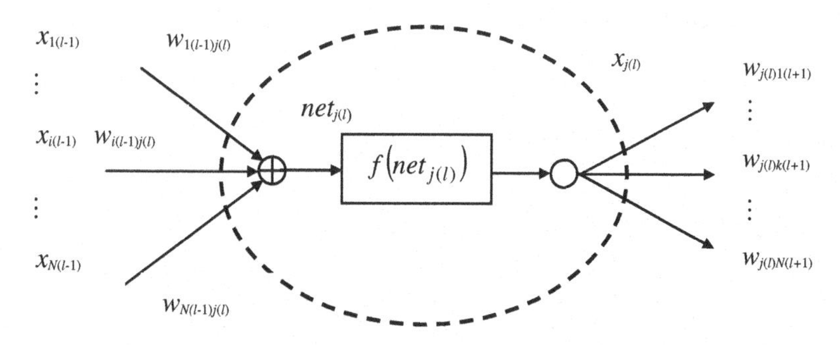

Fig. 7.1 An elementary model of a neuron

M. Krawczak: *Multilayer Neural Networks*, SCI 478, pp. 145–166.
DOI: 10.1007/978-3-319-00248-4_7

Here we consider the differentiable activation functions for generation of outputs from of the neurons. In the model considered an additional element, depicted by O , is included, corresponding to a junction point. Generally, the existence of the junction point in a neuron has been tacitly assumed. Fig. 7.1 shows an extended notation of indices, namely we indicate the position of each neuron in the whole network. For example the weight $w_{i(l-1)j(l)}$ indicates the connection between the neuron $i(l-1)$ belonging to the $(l-1)$-st layer and the neuron $j(l)$ from the l-th layer.

In this way three main elements of a neuron can be distinguished, one element is a summation point, the second is an activation function which transmits the effect of summation, and the third is a junction point spreading the value resulting from the activation function output to neurons of the next layer.

Let us rearrange the neuron's elements in the following way:

- remove the activation function to the outside of the neuron,
- the removed activation functions are shifted to each of the connections between the considered neuron and all neurons of the next layer, becoming thereby the transmittances between neurons,
- the connection between neurons are still weighted,
- the summation point and the junction point make up *a node*,
- the neural network with the rearranged neurons becomes a *flow graph.*

The above rearrangement is pictured in Fig. 7.2.

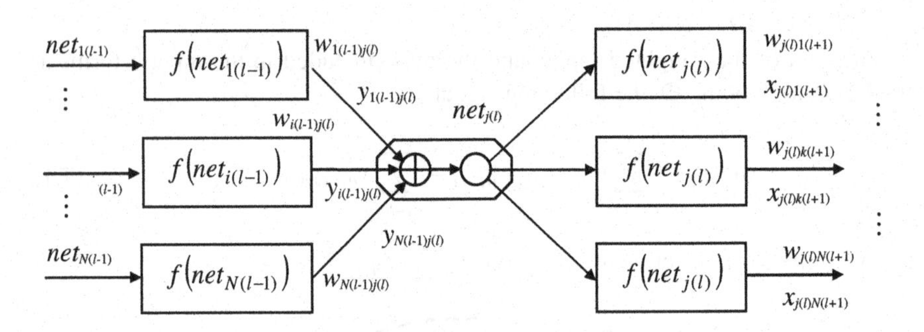

Fig. 7.2 The rearranged neuron with its vicinity as a segment of a flow graph

Now, Equ. 7.1 can be rewritten in the following way

$$
\begin{aligned}
net_{j(l)} &= \sum_{i(l-1)=1}^{N} w_{i(l-1)j(l)}\, x_{i(l-1)} = \\
&\sum_{i(l-1)=1}^{N} w_{i(l-1)j(l)}\, f_{i(l-1)}\left(net_{i(l-1)}\right) = \sum_{i(l-1)=1}^{N} y_{i(l-1)j(l)}
\end{aligned}
\tag{7.2}
$$

and description of any separate edge takes the following form:

$$y_{i(l-1)j(l)} = w_{i(l-1)j(l)}\ x_{i(l-1)} = w_{i(l-1)j(l)}\ f_{i(l-1)}\left(net_{i(l-1)}\right). \tag{7.3}$$

Now let us show an example of a simple neural network with one hidden layer, Fig. 7.3.

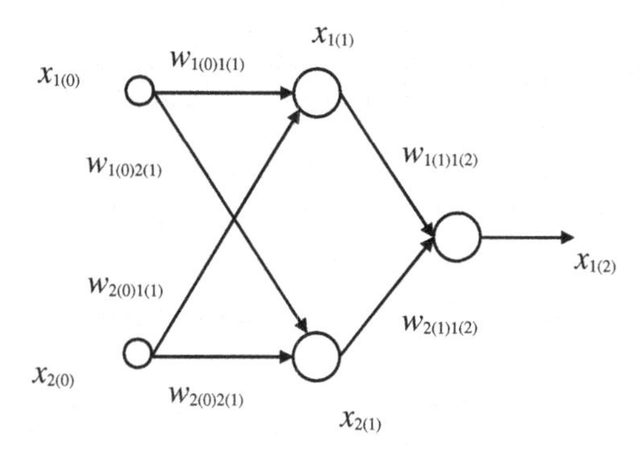

Fig. 7.3 A simple two-layer neural network

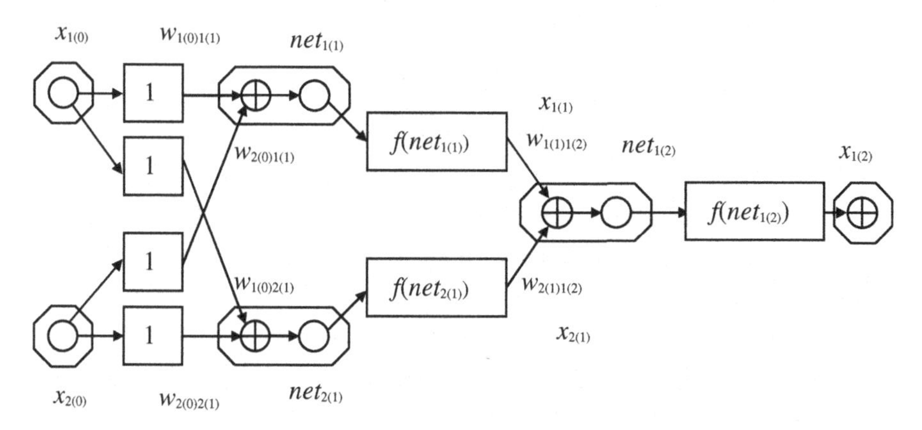

Fig. 7.4 A flow graph corresponding to Fig. 7.3

After using the rearranging procedure described above the same neural network can be considered as a flow graph of Fig. 7.4. When comparing these two pictures we can notice that the architectures of the neural network and the flow graph are exactly the same, in the position of each neuron a graph node has appeared, while to each connection an activation function has been added as a transmittance. In Fig. 7.4 two kinds of reduced nodes appear:

- input nodes are reduced to junction points,
- output nodes are reduced to summation points.

7.2 Elements of the Flow Graph Theory

It has been shown in the previous section how a neural network can be transformed into a flow graph. In this section our attention will be devoted to the basic definitions of reciprocal graphs and the conditions will be established for two mutually transposed graphs to be interreciprocal (Krawczak 2002a, 2003b, 2003d, 2003e, 2005e) and Krawczak and Aladjov (2003).

A graph consists of two types of elements, nodes and edges, and the ways in which these elements are interconnected (Christofides 1975, Campolucci et al. 1997).

Definition 7.1. *A graph is represented by the triple* $G = (V, E, \varphi)$, *where* V *is a nonempty set of "nodes" of the graph,* E *is a set of "edges" of the graph, and* φ *is a mapping from the set* E *to* V *. If the pairs of nodes are connected by an ordered edge (indicated by an arrow), then the edge is "directed". A graph with all edges directed is called "directed graph" or "flow graph". Any two nodes connected by an edge are called "adjacent" nodes. Here, nodes are represented by a small octagon.*

Assumption 7.1. *All nodes consist of two elements, one a summation point and second a junction point (see the following picture)*

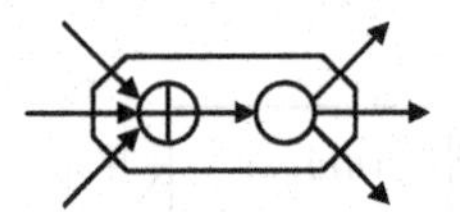

Let us consider a flow graph G with N nodes indexed by i, $i = 1, 2, \ldots, N$. There is a single edge (i, j) between any two nodes i and j, $(i, j = 1, 2, \ldots, N)$; and there is defined a value X_i representing the node; for each pair of nodes a transmittance function T_{ij} is defined; for each edge $(i, j) \in E$ an output is defined, the final value of an edge coming into a node. The above introduced rules are illustrated in Fig. 7.5. The signal passing the transmittance is given by

$$Y_{ij} = T_{ij}\, X_i \,. \tag{7.4}$$

With the Assumption 7.1 we have a relation of a node:

$$X_j = \sum_{i=1}^{N} Y_{ij} + X_{j(0)} = \sum_{i=1}^{N} T_{ij}\, X_i + X_{j(0)} \tag{7.5}$$

where by $X_{j(0)}$ we denote some selected nodes considered as external inputs to the graph, directly feeding the node j (see Fig. 7.5), in this case the associated transmittance is equal to 1.

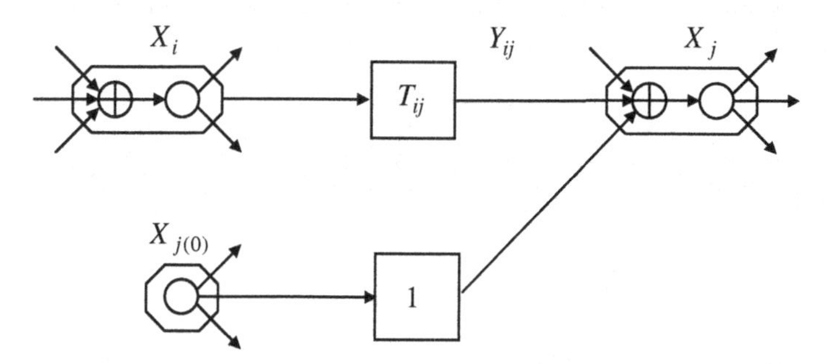

Fig. 7.5 A directed edge of a flow graph

Now we will recall the following definitions and the theorem (Fettweiss 1972, Christofides 1975, Chua and Lin, 1975, Campolucci et al. 1997, Penfield et al. 1979, Shanmugan and Breipohl 1988).

Definition 7.2. *It is said that two graphs* G *and* G^T *are mutually transposed if and only if their transmittances are transposed*

$$T_{ij}^T = T_{ji} \quad \text{for all } i,\ j = 1, 2, \ldots, N\ . \tag{7.6}$$

Definition 5.3: *It is said that two graphs* G *and* G^T *are interreciprocal if and only if*

$$\sum_{j=1}^{N} \left(X_j^T\ X_{j(0)} - X_j\ X_{j(0)}^T \right) = 0\ . \tag{7.7}$$

Theorem 7.1. *Two given transposed flow graphs* G *and* G^T *are interreciprocal.*

Proof: Let us consider a flow graph G and another flow graph G^T that is the transposed graph of G, i.e.

$$T_{ij}^T = T_{ji} \quad \text{for all } \ i,\ j = 1, 2, \ldots, N\ .$$

Now we will transform the following identity

$$\sum_{j=1}^{N} X_j^T\ X_j = \sum_{j=1}^{N} X_j\ X_j^T \tag{7.8}$$

For both considered graphs, using Equ. 7.5 we get

$$X_j = \sum_{i=1}^{N} T_{ij}\ X_i + X_{j(0)} = \sum_{i=1}^{N} Y_{ij} + X_{j(0)}$$

$$X_j^T = \sum_{i=1}^{N} T_{ij}^T\ X_i^T + X_{j(0)}^T = \sum_{i=1}^{N} Y_{ij}' + X_{j(0)}^T$$

and now, by substituting the proper terms in the left- and right-hand sides of (7.9), respectively, we can obtain the following transformation:

$$\sum_{j=1}^{N} X_j^T \left(\sum_{i=1}^{N} T_{ij} X_i + X_{j(0)} \right) = \sum_{j=1}^{N} X_j \left(\sum_{i=1}^{N} T_{ij}^T X_i^T + X_{j(0)}^T \right)$$

$$\sum_{j=1}^{N} X_j^T \left(\sum_{i=1}^{N} Y_{ij} + X_{j(0)} \right) = \sum_{j=1}^{N} X_j \left(\sum_{i=1}^{N} Y_{ij}^T + X_{j(0)}^T \right)$$

$$\sum_{j=1}^{N} \sum_{i=1}^{N} \left(X_j^T Y_{ij} + X_j^T X_{j(0)} \right) = \sum_{j=1}^{N} \sum_{i=1}^{N} \left(X_j Y_{ij}^T + X_j X_{j(0)}^T \right)$$

$$\sum_{j=1}^{N} \sum_{i=1}^{N} \left(X_j^T Y_{ij} - X_j Y_{ij}^T \right) - \sum_{j=1}^{N} \sum_{i=1}^{N} \left(X_j^T X_{j(0)} - X_j X_{j(0)}^T \right) = 0 . \tag{7.9}$$

It was assumed that the graphs G and G^T are mutually transposed, so that the first term of (7.9) is equal zero due to the following reasoning

$$\begin{aligned} \sum_{j=1}^{N} \sum_{i=1}^{N} \left(X_j^T Y_{ij} - X_j Y_{ij}^T \right) &= \sum_{j=1}^{N} \sum_{i=1}^{N} \left(X_j^T T_{ij} X_i - X_j T_{ij}^T X_i^T \right) \\ &= \sum_{j=1}^{N} \sum_{i=1}^{N} \left(X_j^T T_{ij} X_i - X_j T_{ji} X_i^T \right) = 0 . \end{aligned} \tag{7.10}$$

From (7.7) we have that the second term in (7.9) is also equal zero

$$\sum_{i=1}^{N} \sum_{j=1}^{N} \left(X_j^T X_{j(0)} - X_j X_{j(0)}^T \right) = \sum_{j=1}^{N} \left(X_j^T X\, Y_{j(0)} - X_j X_{j(0)}^T \right) = 0 \tag{7.11}$$

which proves that the graphs G and G^T are mutually interreciprocal. □

Here we would like to mention that Equ. 7.9 is known in the electric circuits theory as Tellegen's theorem from 1952 (Penfield et al. 1970, Chua and Lin 1975).

It seems that one case that is of special importance should be distinguished, namely the one of a graph with only one input and only one output. In this case limits Equ. 7.11 are reduced to two terms without the crossing dependencies. In such a simple case we can distinguish a node *in* which is the input to the graph and a node *out* which is the output of the graph, while for the graph G^T a node *in* is the output and a node *out* is the input of the graph. Then, Equ. 7.11 becomes

$$X_{in}\, X_{in}^T = X_{out}^T\, X_{out}\,. \tag{7.12}$$

For the graph G the output X_{out} as a result of the input X_{in} is identical to the output X_{in}^T following the input X_{out}^T for the graph G^T, thus for a single input and a single output the conclusion can be written as

$$\frac{X_{in}}{X_{out}} = \frac{X_{out}^T}{X_{in}^T}\,, \quad X_{in}^T = X_{out}, \quad X_{out}^T = X_{in}\,.$$

Similar interdependence between inputs and output of interreciprocal graphs is valid for multidimensional inputs and outputs, but the one input and one output case illustrates the idea of graph interreciprocity in a very visual way.

7.3 Construction of Adjoint Neural Networks

In the previous section the possibility of conversion of a neural network into a flow graph has been shown, followed by properties of interreciprocal flow graphs.

In this section these properties will allow us to introduce *the adjoint neural networks*. The name "adjoint" has been borrowed from the optimisation as well as optimal control theory, where it is used for additional (adjoint) set of equations, and in the case of the optimal control problems or multistage problems these equations are solved in an opposite direction.

Using the graph theory notation a feedforward network topology can be specified by the following set of equations

$$x_j = \sum_{i=1}^{N} T_{ij}\, w_{ij}\, x_i, \quad j = 1,2,\ldots,N \tag{7.13}$$

where N is the number of all nodes (in our case neurons), x_i, x_j are values describing nodes i and j, $T_{ij} = f_{ij}(\ )$ is a transmittance (or an activation function of neurons e.g., a sigmoid function between the nodes (i and j). Summation is extended over all signals associated with the node x_i, $i = 1,2,\ldots,N$, coming into the node x_j, $j = 1,2,\ldots,N$.

The last equation can be rewritten in a different way by considering the values of nodes, i.e. $net_{j(l)}$, for $l = 1,2,\ldots,L,\ j(l) = 1,2,\ldots,N(l)$, and has one of the forms

$$net_{j(l)} = \sum_{i(l-1)=1}^{N(l-1)} w_{i(l-1)j(1)}\ x_{i(l-1)} = \begin{cases} \sum_{i(0)=1}^{N(0)} w_{i(0)j(1)}\ x_{i(0)}, & \text{if } l = 0 \\ \sum_{i(l-1)=1}^{N(l-1)} w_{i(l-1)j(l)}\ f\left(net_{i(l-1)}\right), & \text{if } l \neq 0 \end{cases} \tag{7.14}$$

but we will specify the following cases depending on the number of the layer

$$net_{j(l)} = \begin{cases} net_{i(0)} = x_{i(0)}, & \text{for } l = 0 \\ \sum_{i(0)=1}^{N(0)} w_{i(0)j(1)}\ x_{i(0)}, & \text{for } l = 1 \\ \sum_{i(l-1)=1}^{N(l-1)} w_{i(l-1)j(1)}\ f\left(net_{i(l-1)}\right), & \text{for } 1 < l < L \\ net_{j(out)} = \sum_{i(L)=1}^{N(L)} w_{i(L-1j(L)}\ f\left(net_{i(L-1)}\right), & \text{for } l = L. \end{cases} \tag{7.14a}$$

Using (7.14a) we can illustrate flows of signals in a neural network treated as a flow graph in the following picture. An example is shown of a two-layer neural network with one hidden layer (Fig. 7.6). A chain of directed edges linking a selected input node $net_{i(0)} = net_{i(in)}$ with a selected output node $net_{i(out)}$ is presented in the figure.

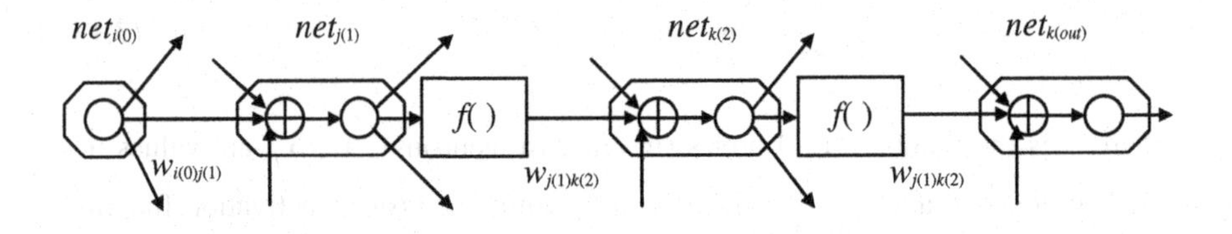

Fig. 7.6 A schematic exemplary two-layer neural network as a flow graph

Now let us remind the learning process of the neural networks. Learning of the neural networks consists in changing the weights when the desired output $d_{pj(L)}$, $p = 1, 2, \ldots, P$, and the actual output $x_{pj(L)}$, resulting both from the input $x_{i(0)p}$, $i(0) = 1, 2, \ldots, N(0), j(L) = 1, 2, \ldots, N(L)$, are different. The index p is a training example, while L denotes the number of layers in the network. The change is done by gradient descent,

$$\Delta w_{i(l-1)j(l)} = w^{new}_{i(l-1)j(l)} - w^{old}_{i(l-1)j(l)} = -\eta \frac{\partial E}{\partial w_{i(l-1)j(l)}} \tag{7.15}$$

where η is the learning rate, and E is the learning performance. Generally, the learning performance E is defined as the sum over all the training pattern examples

$$E = \sum_{p=1}^{P} \sum_{j(L)=1}^{N(L)} E_{pj(L)} \left(d_{pj(L)}, x_{pj(L)}\right) \tag{7.16}$$

where $N(L)$ is a number of output nodes (neurons). For a specific training pattern p we use the squared difference between the patterns and the actual network output

$$E_p = \frac{1}{2} \sum_{j(L)=1}^{N(L)} \left(d_{p(L)} - x_{pj(L)}\right)^2. \tag{7.17}$$

Using directly the delta rule, derived in Chap. 4, for updating the weights $w_{i(l-1)j(l)}$, we obtain

$$\Delta w_{i(l-1)j(l)} = -\eta \frac{\partial E_p}{\partial w_{i(l-1)j(l)}} = -\eta \frac{\partial E_p}{\partial net_{j(l)}} \frac{\partial net_{j(l)}}{\partial w_{i(l-1)j(l)}} = \eta \delta_{j(l)} x_{i(l-1)} \tag{7.18}$$

where $j(l) = 1, 2, \ldots, N(l), \quad i(l-1) = 1, 2, \ldots, N(l-1), \quad l = 1, 2, \ldots, L$.

Let us recall the expression for δ in (7.18)

$$\delta_{i(l-1)} = \begin{cases} \sum\limits_{j(L)=1}^{N(L)} \left(d_{pj(L)} - x_{pj(L)}\right) f'\left(net_{j(L)}\right), & \text{if } l = L \\ f'\left(net_{i(l-1)}\right) \sum\limits_{j(l)=1}^{N(l)} w_{i(l-1)j(l)}\, \delta_{j(l)}, & \text{if } l \neq L \end{cases} \tag{7.19}$$

$$\delta_{j(out)} = \left(d_{pj(L)} - x_{pj(L)}\right) = e_{j(L)}$$

$$\delta_{i(l-1)} = \begin{cases} \sum\limits_{j(L)=1}^{N(L)} f'\left(net_{j(L)}\right) \delta_{j(out)}, & \text{if } l = L \\ f'\left(net_{i(l-1)}\right) \sum\limits_{j(l)=1}^{N(l)} w_{i(l-1)j(l)}\, \delta_{j(l)}, & \text{if } 2 \leq l \leq L-1 \\ \delta_{i(0)} = \delta_{i(in)} = \sum\limits_{j(0)=1}^{N(0)} w_{i(0)j(1)}\, \delta_{j(1)}, & \text{if } l = 1. \end{cases} \tag{7.19a}$$

It can be easily noticed that Equ. 7.14 and 7.19 have the same structure. In (7.14) the signals *net* flow from the inputs through the network to the outputs, while in (7.19) the signals δ flow in the opposite direction from the outputs to the inputs. In (7.19) the influence of a performance index of learning is included.

According to the definition of interreciprocity of flow graphs it is required to define the input to the adjoint graph. The term $\delta_{j(out)} = \left(d_{pj(L)} - x_{pj(L)}\right) = e_{j(L)}$ depends on the shape of the performance index, and can be treated as the input to the adjoint network. In the original networks, the input nodes are distinguished, and the response of the network is simple $\delta_{j(out)} = e_{j(L)}$. In order to demonstrate some similarity we must consider input nodes of the original network as output nodes of the adjoint network, and vice versa - the output nodes of the original network as input nodes of the adjoint network, equal to $e_{j(L)}$.

Now we must determine the form of the transmittances of the reciprocal graph, i.e.

$$T_{ij}^{T} = T_{ji}, \quad \text{for all} \quad i,\ j = 1,2,\ldots,N \tag{7.20}$$

For the normal signals flow direction we can write the following relationship

$$net_{j(l)} = T_{i(l-1)j(l)}\, w_{i(l-1)j(l)}\, net_{i(l-1)} = f(net_{i(l-1)})\, w_{i(l-1)j(l)}\, net_{i(l-1)} \tag{7.21}$$

while for the opposite signal flow direction we can write

$$\delta_{i(l-1)} = S_{j(l)i(l-1)}\, w_{i(l-1)j(l)}\, \delta_{j(l)} \tag{7.22}$$

where $S_{j(l)i(l-1)}$ is the transmittance for the delta signals. Let us differentiate (7.21) and rewrite (7.22) using the definition of the delta rule. We obtain

$$\frac{\partial net_{j(l)}}{\partial net_{i(l-1)}} = \frac{\partial f(net_{i(l-1)})}{\partial net_{i(l-1)}}\, w_{i(l-1)j(l)} \tag{7.23}$$

$$\begin{aligned} \frac{\partial E}{\partial net_{i(l-1)}} &= S_{j(l)i(l-1)}\, w_{i(l-1)j(l)} \frac{\partial E}{\partial net_{j(l)}} \\ \frac{\partial net_{j(l)}}{\partial net_{i(l-1)}} &= S_{j(l)i(l-1)}\, w_{i(l-1)j(l)} \frac{\partial E}{\partial E} \end{aligned} \tag{7.24}$$

and by substituting (7.23) into (7.24) we get

$$S_{j(l)i(l-1)} = T_{j(l)i(l-1)}^{T} = \frac{\partial f(net_{i(l-1)})}{\partial net_{i(l-1)}} = f'\left(net_{i(l-1)}\right). \tag{7.25}$$

In this way we have shown the form of the transposed transmittances appearing in the adjoint neural networks. In Fig. 7.6 we have presented an exemplary path of a two-layer neural network, while a counterpart to this example - a path of the adjoint neural network is shown in Fig. 7.7.

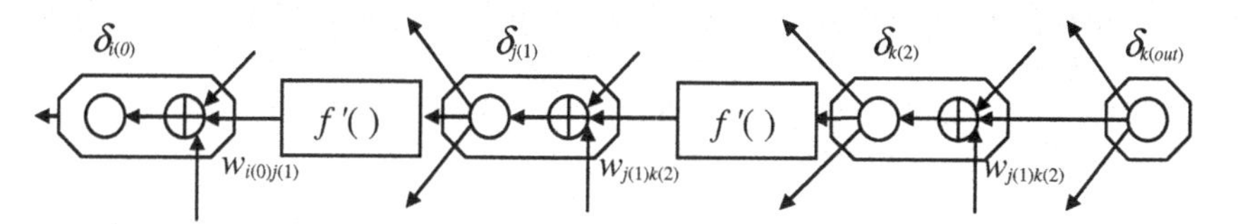

Fig. 7.7 A schematic exemplary two-layer neural network as an adjoint neural network

Looking at Equ. 7.14 and 7.19 we can notice that both networks are topologically identical, meaning that there is a strict and unique correspondence between respective signals and connections. The adjoint network is found by application of the same network architecture, reversal of the direction of signal flows, replacement of activation functions by their derivatives, and switching of the positions of summing points with junction points within each node.

It is easy to notice that the transformation of the original network into the adjoint network is governed by very simple rules that will be described in the next section.

In the transposed graph the signals flow in the reverse direction to that of the original graph G and by virtue of the transposed graph becomes *the adjoint network*. The adjoint network is characterized by the inputs (which are the outputs of the original network) and the outputs (which are the inputs to the original network).

7.4 Neural Networks versus Networks

In some works of the present author, e.g. (Krawczak 2000a), the multilayer neural networks were divided into two kinds of layers. The first kind consisted only of weights with summation points while the second only of activation functions. The idea was just the same as in this volume, the simple weight layers bring related to nodes and the activation function layers being related to edges with transmittances.

When considering the architecture of any class of neural networks, e.g. feedforward networks, we can observe four basic building elements of networks:

- summing points,
- junction points,
- univariate functions,
- multivariate functions (Krawczak 2000a).

Any architecture of neural networks can be represented as a flow graph by introduction of the following changes:

- a neuron is divided into three parts: a summation point, a junction point and an activation function,
- a summation point together with a junction point becomes a node of a flow graph (of a network),
- an activation function becomes a transmittance of each outgoing edge,
- a transmittance multiplied by a synaptic weight.

In the previous section we have shown the idea of changing a neural network into a network, and next deriving the adjoint network. Construction of an adjoint network requires the reversal of the flow direction in the original network, the labelling of all resulting signals as δ_j, and performing of the following operations:

1. *A summation point and junction point are put together as a node*

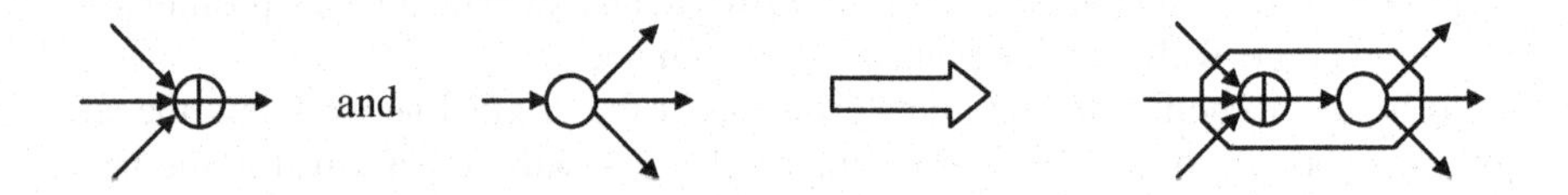

2. *Univariate functions are replaced by their derivatives*

$x_{i(l-1)}$ $\quad f(net_{j(l-1)})$ $\quad x_{j(l)}$ $\quad w_{i(l-1)j(l)}$ $\quad \Rightarrow \quad$ $\delta_{i(l-1)}$ $\quad f'(net_{j(l-1)})$ $\quad \delta_{j(l)}$ $\quad w_{i(l-1)j(l)}$

Explicitly, a continuous scalar function $x_j = f(w_{ij}, x_i)$ is replaced by $\delta_i = f'(w_{ij}, x_i)\delta_j$, where $f'(w_{ij}, x_i) = \partial x_j / \partial x_i$. Special cases are

- *weights:* $x_j = w_{ij}\, x_i$, in whose case $\delta_i = w_{ij}\, \delta_j$

$x_{i(l-1)}$ $\quad x_{j(l)}$ $\quad w_{i(l-1)j(l)}$ $\quad \Rightarrow \quad$ $\delta_{i(l-1)}$ $\quad \delta_{j(l)}$ $\quad w_{i(l-1)j(l)}$

- *bipolar activation function:* for the function $x_j = \tanh(x_i)$ the derivative is $f'(x_i) = (1 - x_j^2)$

$x_{i(l-1)} \rightarrow \boxed{\tanh(x_i)} \rightarrow x_{j(l)} \quad \Longrightarrow \quad \delta_{i(l-1)} \leftarrow \boxed{\left(1-x_j^2\right)} \leftarrow \delta_{j(l)}$

- *unipolar activation function:* for the function $x_j = \dfrac{1}{1+\exp(-x_i)}$ the derivative is $f'(x_i) = (1-x_j)$

$x_{i(l-1)} \rightarrow \boxed{\dfrac{1}{[1+\exp(-x_i)]}} \rightarrow x_{j(l)} \quad \Longrightarrow \quad \delta_{i(l-1)} \leftarrow \boxed{x_i\left(1-x_i^2\right)} \leftarrow \delta_{j(l)}$

3. *Multivariate functions are replaced with their Jacobians*

$x_{in} \rightarrow \boxed{F(x_{in})} \rightarrow y_{out} \quad \Longrightarrow \quad \delta_{in} \leftarrow \boxed{\partial F/\partial x_{in}} \leftarrow \delta_{out}$

A multivariate function maps a vector of input signals into a vector of output signals, $y_{out} = F(x_{in})$. In the transformed network, the derivative is replaced by the Jacobian $\partial F(x_{in})/\partial x_{in}$ (a matrix of partial derivatives), and we have $\delta_{in} = \partial F(x_{in})/\partial x_{in}\, \delta_{out}$.

4. *Outputs become inputs*

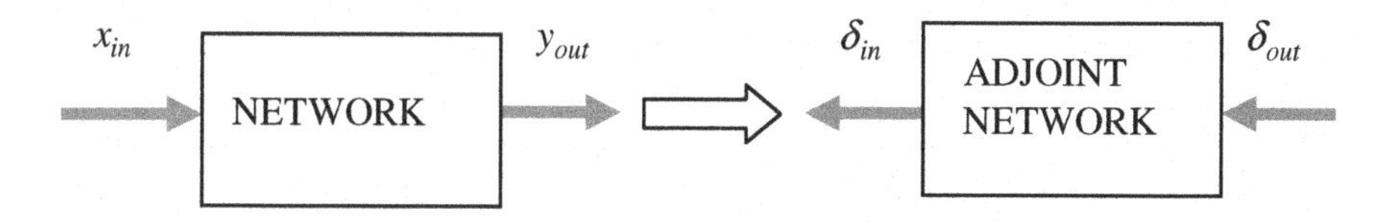

By reversing the signal flow, output nodes y_{out} in the considered network become input nodes in the adjoint network. These inputs are then set taken to be e. For cost functions different than the squared error considered here, the input should be set to $\partial E / \partial y_{out}$.

The considered three rules allow us for a simple construction of the adjoint network from the original network. Note that there is a topological equivalence between the two networks. The order of computations in the adjoint network is thus identical to the order of computations in the original network. The signals

δ_j that propagate through the adjoint network correspond to the terms $\partial E/\partial x_j$ necessary for gradient adaptation. The exact equations can be taken from the adjoint network, which would complete the derivation.

7.5 Derivation of the Backpropagation Algorithm

The simplicity of application of the adjoint neural networks in deriving the error backpropagation algorithm will be clarified in this section. We derive the standard backpropagation algorithm. For the sake of consistency with the traditional notation, we have labelled the summation signal $net_{i(l)}$ with the enlarged subscript to denote the layer, see Fig. 7.5. The adjoint network shown in Fig. 7.6 is found by applying the construction rules described above. From this figure we may immediately write down the equations for calculating the delta terms, Equ. 7.19

$$\delta_{j(out)} = \left(d_{j(L)p} - x_{j(L)p}\right) = e_{j(L)}$$

$$\delta_{i(l-1)} = \begin{cases} \sum_{j(L)=1}^{N(L)} f'\left(net_{j(L)}\right)\delta_{j(out)}, & \text{if } l = L \\ f'\left(net_{i(l-1)}\right) \sum_{j(l)=1}^{N(l)} w_{i(l-1)j(l)}\, \delta_{j(l)}, & \text{if } l \neq L. \end{cases}$$

Equ. 7.18. for the weight update now can be formulated as $\Delta w_{i(l-1)j(l)} = \eta \delta_{j(l)} x_{i(l-1)}$.

These two equations, (7.14) and (7.19), precisely describe the standard backpropagation, the equations well known from many textbooks about feedforward neural networks. There is no doubt that the presented approach gives a much more easy way to obtain the equations describing the backpropagation algorithm.

In these sections we have demonstrated the way for the derivation of the gradient-based algorithms for any network architectures. For any interesting neural network we can construct the adjoint neural network. The adjoint network is built using three rules of changing the original network, and can be obtained in a very simple way. All formulae for the learning algorithms can be derived using this methodology.

7.6 Description of the Adjoint Neural Networks

In this section we will introduce the generalized net description of the adjoint neural network for the backpropagation algorithm (Krawczak 2002a, Krawczak and Aladjov 2002, Krawczak 2003c,2003d, 2003e).

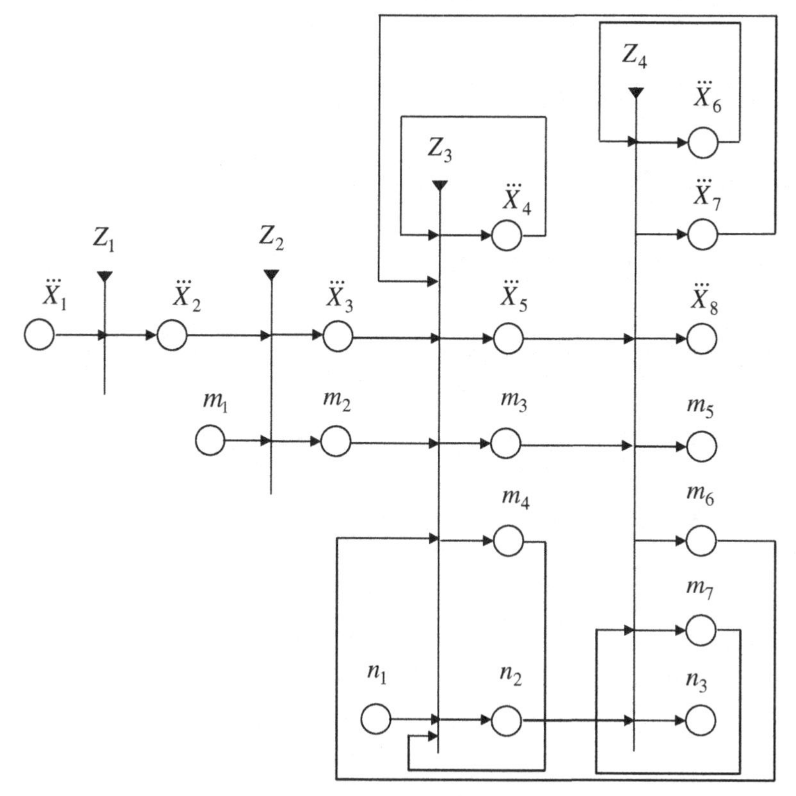

Fig. 7.8 The generalized net representation of the adjoint neural network

The generalized net representation of the adjoint neural networks contains four transitions, see Fig. 7.8. Each transition represents a separate stage of the adjoint neural network functioning. These stages are as follows:

- construction of the adjoint neural network,
- initialisation of connection weights,
- propagation of signals,
- back propagation of the error.

In this section we will constrain our consideration to some elements of the reduced generalized net form, in order to show how this methodology can be used to construct the description.

In the considered multilayer neural network each neuron is represented by a single neuron of α -type, and the token has the following initial characteristic

$$y(\alpha_I)=\langle NN1,\ l, I,\ f_I, x_I,\ imW,\ d\rangle,\quad \text{for}\ \ l=0,1,\ldots,L \tag{7.26}$$

where

$NN1$
the neural network identifier,

l
the layer number,

I
the number of the token (neuron),

f_I
an activation function of the I -th neuron,

x_I
the current value of neuron output,

imW
the index matrix of the weights, which contains the connection, having the following form

$$imW = \begin{array}{c|cc} & in & out \\ \hline 1 & W_{1,I} & W_{I,1} \\ 2 & W_{2,I} & W_{I,2} \\ \vdots & \vdots & \vdots \\ N & W_{N,I} & W_{I,N} \end{array} \tag{7.27}$$

where

N
the number of all neurons in the considered neural network,

$W_{m,n}$
the weight connecting the m -th neuron with the n -th neuron,

d
description, which can be defined as follows

$$d = \begin{cases} "in" \text{ if } (\forall i \in (1,2,...,N))\,(imW_{i,I} = 0) \& (\exists i \in (1,2,...,N)(imW_{I,i} \neq 0)) \\ "out" \text{ if } (\forall i \in (1,2,...,N))(imW_{I,i} = 0) \& (\exists i \in (1,2,...,N)(imW_{i,I} \neq 0)) \\ "int" \text{ if } (\exists i \in (1,2,...,N))(imW_{I,i} \neq 0) \& (\exists i \in (1,2,...,N)(imW_{i,I} \neq 0)) \\ "iso" \text{ if } (\forall i \in (1,2,...,N))(imW_{I,i} = 0) \& (\forall i \in (1,2,...,N)(imW_{i,I} = 0)) \end{cases} \tag{7.28}$$

where "*in*", "*out*", "*int*", "*iso*" denote the input, output, internal and isolated neurons, respectively.

It is worth noticing that the characteristic (7.26) includes all information required to estimate the whole neural network, namely the connectivity and characteristics.

The process of the adjoint neural network construction is based on changes of the neurons features, that is, the neurons must be able to propagate the signals in the forward direction as well as to propagate the error in the back direction, and to possess all the information required for the connection weights evaluation. These changes of the new neuron features are represented by generation of the new characteristics of the tokens in the place $\dddot{X}_2$, which are as follows

$$y(\alpha_I) = \langle NN1, l, I, f_I, f_I', x_I, \delta_I, imW, imW_1, d, d_1 \rangle \tag{7.29}$$

where the new components of the characteristics have the following meaning

$$f_I'(net_I) = \frac{\partial f_I(net_I)}{\partial net_I} \tag{7.30}$$

where net_I is related to Equ. 7.14, and

$$\delta_I = -\frac{\partial E}{\partial net_I} \quad \text{(related to Equ. 7.19)} \tag{7.31}$$

imW_1

has the same components as imW, but the weights of inputs are replaced by the weights of outputs and vice versa, that is

$$imW_1 = \begin{array}{c|cc} & in & out \\ \hline 1 & W_{I,1} & W_{1,I} \\ 2 & W_{I,2} & W_{2,I} \\ \vdots & \vdots & \vdots \\ N & W_{I,N} & W_{N,I} \end{array} \tag{7.32}$$

d_1

describes the connectivity of the neurons within the adjoint neural network, and can be obtained from d in the following way

$$d_1 = \begin{cases} "in" & \text{if } d = "out" \\ "out" & \text{if } d = "in" \\ "int" & \text{if } d = "int" \\ "iso" & \text{if } d = "iso". \end{cases} \tag{7.33}$$

This transformation can be represented by construction of the new adjoint neurons, and next the adjoint neural network, which has exactly the same structure as the considered neural network. Following the consideration of the previous sections the model of a neuron has the form depicted in Fig. 7.9.

Let us construct a new processing element, which consists of two parts, the first part is the modified neuron model (Fig. 7.9), and the second part is the adjoint neuron model (related to Figure 7.7), and is shown in Fig. 7.10.

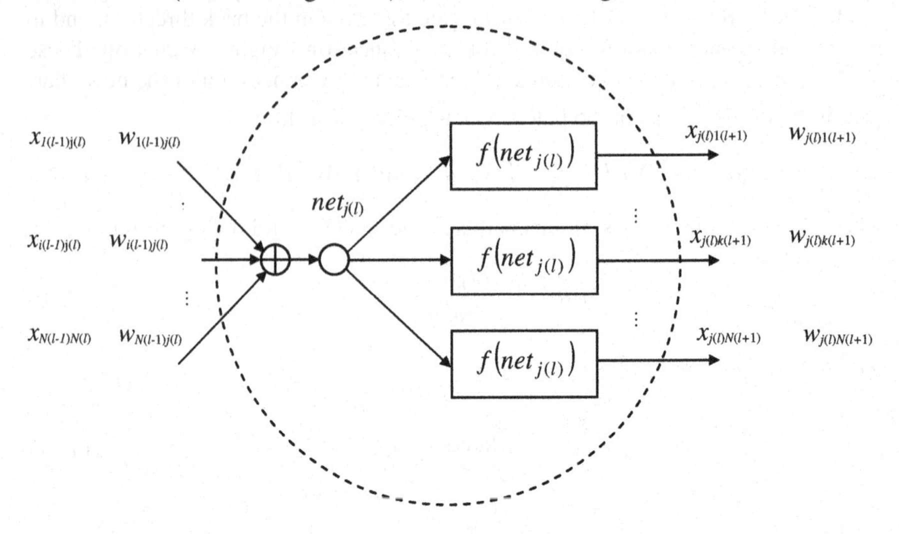

Fig. 7.9 The modified neuron model

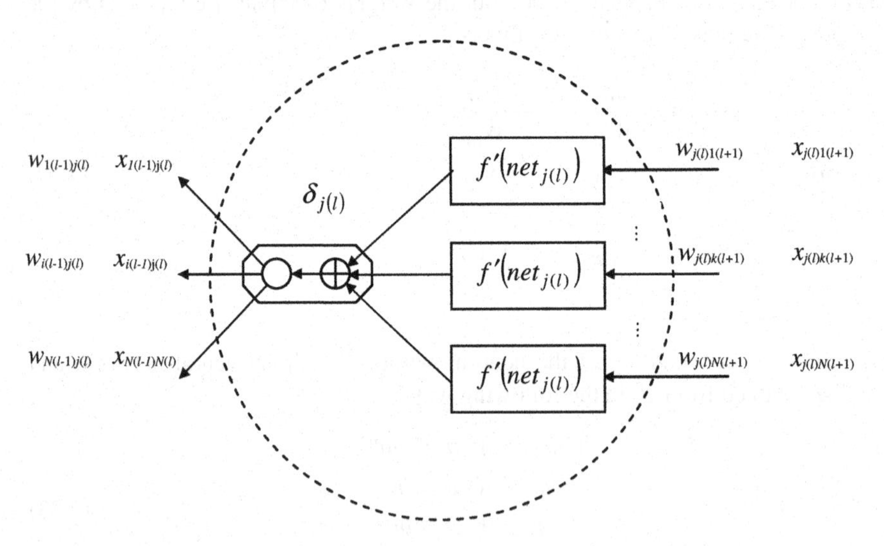

Fig. 7.10 The model of the adjoint neuron

The new elements are arranged in a structure of the considered neural network. In detail the signals in the new processing elements are described as follows

$$f'_{j(l)} = -\frac{\partial E}{\partial net_{j(l)}} \frac{\partial f_{j(l)}\left(\sum_{i(l-1)=1}^{n} w_{i(l-1)j(l)}\, x_{i(l-1)}\right)}{\partial net_{j(l)}}$$

$$net_{j(l)} = \begin{cases} net_{i(0)} = x_{i(0)}, & \text{for } l = 0 \\ \sum_{i(0)=1}^{N(0)} w_{i(0)j(1)}\, x_{i(0)}, & \text{for } l = 1 \\ \sum_{i(l-1)=1}^{N(l-1)} w_{i(l-1)j(1)}\, f\left(net_{i(l-1)}\right), & \text{for } 1 < l < L \\ net_{j(out)} = \sum_{i(L)=1}^{N(L)} w_{i(L-1j(L)}\, f\left(net_{i(L-1)}\right), & \text{for } l = L \end{cases}$$

$$\delta_{i(l-1)} = \begin{cases} \sum_{j(L)=1}^{N(L)} f'\left(net_{j(L)}\right)\delta_{j(out)}, & \text{if } l = L \\ f'\left(net_{i(l-1)}\right) \sum_{j(l)=1}^{N(l)} w_{i(l-1)j(l)}\, \delta_{j(l)}, & \text{if } 2 \le l \le L-1 \\ \delta_{i(0)} = \delta_{i(in)} = \sum_{j(0)=1}^{N(0)} w_{i(0)j(1)}\, \delta_{j(1)}, & \text{if } l = 1 \end{cases}$$

where $j(l) = 1, 2, \ldots, N(l)$, $i(l-1) = 1, 2, \ldots, N(l-1)$, $l = 1, 2, \ldots, L$.

From the generalized net point of view, the process of the adjoint neural network can be represented by the following transition Z_1

$$Z_1 = \langle \{\dddot{X}_1\}, \{\dddot{X}_2\}, \begin{array}{c|c} & \dddot{X}_2 \\ \hline \dddot{X}_1 & true \end{array} \rangle. \tag{7.34}$$

The next transition Z_2 describes the first stage of the training process, namely the initialisation of weights

$$Z_2 = \langle \{\dddot{X}_2, m_1\}, \{\dddot{X}_3, m_2\}, \begin{array}{c|cc} & \dddot{X}_3 & m_2 \\ \hline \dddot{X}_2 & true & false \\ m_1 & false & true \end{array} \rangle. \quad (7.35)$$

In the place m_1 a token associated with the performance index enters the generalized net with the following initial characteristic

$$y(\beta) = \langle NN1, E, E_{\max} \rangle \quad (7.36)$$

where

$NN1$
the neural network identifier,

E
the performance index of the neural network learning,

$E_{\max}$
the threshold value of the performance index, which must be reached.

In the place m_2 this token does not change this characteristic.

The transition Z_3 associated with the pattern recognition process has the form

$$Z_3 = \langle \{\dddot{X}_3, \dddot{X}_4, \dddot{X}_7, m_2, m_6, n_1\}, \{\dddot{X}_4, \dddot{X}_5, m_3, m_4, n_2\}, \begin{array}{c|ccccc} & \dddot{X}_5 & \dddot{X}_6 & m_3 & m_4 & n_2 \\ \hline \dddot{X}_3 & V_1 & false & false & false & false \\ \dddot{X}_4 & false & V_2 & false & false & false \\ \dddot{X}_7 & V_1 & false & false & false & false \\ m_2 & false & false & V_3 & \neg V_3 & false \\ m_4 & false & false & V_3 & \neg V_3 & false \\ m_6 & false & false & V_3 & \neg V_3 & false \\ n_1 & false & false & false & false & V_3 \end{array} \rangle \quad (7.37)$$

where

V_1 = "the neuron has assigned the input signal",

V_2 = "the neuron has assigned the output signal",

V_3 = "all neurons of the network have assigned the output signals".

The tokens γ_p, $p=1,2,\ldots,P$, where p is the number of the training pattern enter the place n_1 with the initial characteristic

$$y(\gamma_p)=\langle X_p(0), D_p, p\rangle \tag{7.38}$$

where $X_p(0)$ - is the inputs vector of the neural network, and D_p is the vector of desired network outputs.

The tokens of α -type enter the place $\dddot{X}_4$ for the purpose of calculation of the neuron outputs, and next the tokens are transferred to the place $\dddot{X}_5$, where the tokens α form the output layer (for $d =$"out"), and obtain the new characteristics in the following form

$$y(\alpha_I)=\langle NN1, l, I, p, f_I, f_I', x_I, \bar{\delta}_I, im\overline{W}, im\overline{W}_1, d, d_1\rangle \tag{7.39}$$

related to the nominal values of connection weights and the pattern p. The factor $\bar{\delta}_I$ can be viewed as representing the inputs for the adjoint neural network.

In the place m_3 the token β obtains the new characteristic

$$y(\beta)=\langle NN1, E', E_{\max}\rangle. \tag{7.40}$$

The next transition Z_4 is responsible for the error propagation via the adjoint neural network, and the weights correction process and has the form

$Z_4 = \langle\{\dddot{X}_5, \dddot{X}_6, m_3, n_2, m_7\}, \{\dddot{X}_6, \dddot{X}_7, \dddot{X}_8, m_5, m_6, m_7, n_3\},$

	$\dddot{X}_6$	$\dddot{X}_7$	$\dddot{X}_8$	m_5	m_6	m_7	n_3
$\dddot{X}_5$	V_4	$false$	$false$	$false$	$false$	$false$	$false$
$\dddot{X}_6$	$false$	V_5	$\neg V_5$	$false$	$false$	$false$	$false$
m_3	$false$	$false$	$false$	$\neg V_5$	$V_5 \& \neg V_6$	V_6	$false$
m_7	$false$	$false$	$false$	$\neg V_5$	$V_5 \& \neg V_6$	V_6	$false$
n_2	$false$	$false$	$false$	$false$	$false$	$false$	$true$

$\rangle$ (7.41)

where

$V_4 =$ "the adjoint neuron has assigned the input signal",

$V_5 =$ "the new pattern for learning must be applied",

$V_6 =$ "there exist adjoint neurons with unassigned outputs".

In the place $\dddot{X}_6$ the new values of the parameters δ_l are assigned to the adjoint neuron outputs, and the weights can be corrected, in details according to the following rule

$$\Delta w_{i(l-1)j(l)} = \eta \delta_{j(l)} x_{i(l-1)} .$$

Other details of the transitions, places and tokens are highly similar to those considered in the previous sections, and we will not repeat them.

The process of the backpropagation learning algorithm of the multilayer neural networks can be simplified by introducing the adjoint neurons. Such neurons enrich the ordinary neuron capabilities with some mechanism for error backpropagation and self-modification of the connection weight changes. It seems that the new structure of the neurons (the neuron combined with the adjoint neuron) gives the possibilities for hardware implementation of the neural network including the mechanism of the backpropagation learning algorithm or other learning algorithms based on gradient descent.

Chapter 8
Summary

In this book, the natural structure of multilayer neural networks was used in order to consider this class of neural networks from the systems point of view: the aggregated neurons lying within one layer constitute the stage of the system and the bordering stages exchange the information on their states. The outputs of the neurons, from the same layer, constitute the state vectors. Connection weights between the neurons of the same layer are arranged in vectors and are treated as controls. In this way, we developed the interpretation of the multilayer neural networks as the multistage control systems.

The problem of connection weights adjustment or the problem of learning of multilayer neural networks was converted into the particular problem of optimal control of the multistage systems.

A new class of learning algorithms was developed, which is conceptually based on dynamic programming. The only way to avoid the computational burden in training the neural networks, which are large scale systems, is to introduce some approximation of the return functions appearing in the dynamic programming methodology.

The first new learning algorithm considers the first order approximation of the Taylor expansion of the return functions – the algorithm is called the first order differential dynamic programming algorithm. In the backpropagation algorithm, all weights are changed in order to improve the performance index, while in the case of the first order differential dynamic programming algorithm, the changes of the weights are performed until the minimum of the return function, associated with each stage, is reached; after that the computation is shifted to another layer.

The second order differential dynamic programming algorithm was also developed, but in this case it must be assumed that the appearing Hessian matrices are nonsingular, and the algorithm requires computation of the inverse matrices, which creates many computational problems.

The gain parameter was introduced to the sigmoidal neuron models. The changes of the value of this parameter allow to "linearize", in some sense, the neuron activation functions. For a small value of this parameter the "almost" linear-quadratic optimal control problems were developed. Two new parameterised learning algorithms, based on the first order differential dynamic programming algorithm, coordinating the changes of the gain parameter, were developed. In the

M. Krawczak: *Multilayer Neural Networks*, SCI 478, pp. 167–169.
DOI: 10.1007/978-3-319-00248-4_8

first procedure, borrowed from the continuation method, the value of the gain parameter gradually increased within the predefined range. The second procedure has additional elements of the dynamic programming methodology and allows for finding the optimal value of the gain parameter. The developed algorithm is called the heuristic dynamic programming algorithm.

The learning of the multilayer neural networks was also considered as a multiobjective optimisation problem. Consequently, the updating algorithm in the learning process of the neural networks was developed. For the already trained neural networks, the approach gives the possibility of updating the connection weights in the case when new training example must be involved, and we do not need to train the network from the beginning, which saves time of the training process.

We introduced and developed the adjoint neural networks. The approach was based on the graph theory, allowing for the introduction of the reciprocal graphs in which signals flow in the opposite direction. The formulae for signals in the adjoint neural networks represent the known formulae from the backpropagation algorithm, which was considered in details. The approach can be extended to other gradient based learning algorithms for neural networks, also for the first order differential dynamic programming algorithm, the parameterised first order differential dynamic programming algorithm, and for the heuristic dynamic programming algorithm. It must be emphasised that the adjoint neural networks provide the basis for constructing the neural networks and they can be used for the hardware implementation of the learning algorithm.

The above listed results are spread throughout the book and can be treated as an excuse to introduce the generalized net methodology for new representation of discrete event dynamic systems. The generalized net methodology is exploited in order to construct new representations of neural networks functioning. The new descriptions show the dynamics as well as logical organisation of the neural networks as systems. Different aggregations show different roles of information carried by tokens and the simplicity of the third representation is accompanied by the increase in complexity of the token characteristics.

Much attention was paid to the development of the generalized net representations of the new learning algorithms based on dynamic programming. For the generalized net description of the classic backpropagation algorithm three types of tokens were introduced: one for representation of neurons, the second - related to the performance index of learning, and the last representing the training patterns. In the same way, the generalized net representation of the first order differential dynamic programming algorithm was developed. It is interesting that the structures of these two representations are very similar. The generalized net description of the heuristic dynamic programming algorithm showed a very slight change of the previous two representations; the extra transition is responsible for the optimal adjustment of the gain parameter. It seems that modelling of other learning algorithms will only require very small modifications of the generalized net description of the standard backpropagation algorithm.

The introduced adjoint neural networks give the formulae for any gradient based learning algorithm in a very simple and direct way. Therefore, the generalized net description of the adjoint neural networks has a different structure than that described above. The developed generalized net representation of the adjoint neural networks is logically separated in three parts, representing different functions: the changes of the neuron properties, the properties of the learning algorithm, and the information about training patterns. It seems that this model has some similarities with analogue computers used many years ago for modelling of systems, therefore, the model can be used for hardware implementation of multilayer neural networks with built in any gradient based learning algorithm. Such implementation will allow for training the neural networks in short time.

In this book, the concept of the generalized net was used for representing the functioning of multilayer neural networks, and numerous sophisticated tools were used in an informal manner, like the operators, which have a major theoretical and practical value because they allow studying the properties and the behaviour of the considered neural network processes.

The generalized net approach to modelling of real systems may be used successfully for the description of a variety of technological and intellectual problems, it can be used not only for representing the parallel functioning of homogenous objects, but also for modelling non-homogenous systems, for example systems which consist of a different kind of subsystems.

References

1. Aarts, E., Laarhoven, P.: Simulated Annealing: Theory and Practice. Wiley and Sons, New York (1987)
2. Amari, S.-I.: A universal theorem on learning curves. Neural Networks 6, 161–166 (1993)
3. Anderson, J.A., Rosenfeld, E.: Neurocomputing: Foundation of Research. MIT Press, Cambridge (1989)
4. Atanassov, K.: On the concept "generalized net". AMSE Review 1(3), 39–48 (1984)
5. Atanassov, K.: Generalized nets and their fuzzings. AMSE Review 2(3), 39–49 (1985)
6. Atanassov, K.: Generalized index matrices. Competes Rendus de l'Academie Bulgare des Sciences 40(11), 15–18 (1987)
7. Atanassov, K.: Generalized Nets. World Scientific, Singapore (1991)
8. Atanassov, K.: Introduction to the Theory of the Generalized Nets. Pontica Print, Bourgas (1992) (in Bulgarian)
9. Atanassov, K. (ed.): Applications of Generalized Net. World Scientific, Singapore (1993)
10. Atanassov, K.: Intuitionistic Fuzzy Systems. BUSEFAL 58, 92–96 (1994)
11. Atanassov, K.: Generalized Nets and Systems Theory. Prof. M. Drinov Academic Publishing House, Sofia (1997)
12. Atanassov, K.: Generalized Nets in Artificial Intelligence. Generalized Nets and Expert Systems, vol. 1. Prof. M. Drinov Academic Publishing House, Sofia (1998)
13. Atanassov, K.: Algebraic aspect of the generalized net theory. Issues in Intuitionistic Fuzzy Sets and Generalized Nets 5, 1–9 (2005)
14. Atanassov, K.: On a new hierarchical operator over the generalized nets. Issues in Intuitionistic Fuzzy Sets and Generalized Nets 3, 29–34 (2006)
15. Atanassov, K.: On Generalized Nets Theory. Prof. M. Drinov Academic Publishing House, Sofia (2007)
16. Atanassov, K., Aladjov, H.: Generalized Nets in Artificial Intelligence. Generalized Nets and Machine Learning, vol. 2. Prof. M. Drinov Academic Publishing House, Sofia (2000)
17. Atanassov, K., Baczyński, M., Drewniak, J., Kacprzyk, J., Krawczak, M., Szmidt, E., Wygralak, M., Zadrożny, S. (eds.): Developments in fuzzy sets, intuitionistic fuzzy sets, generalized nets and related topics. Foundations, vol. I. SRI Polish Academy of Sciences, Warsaw (2010)
18. Atanassov, K., Baczyński, M., Drewniak, J., Kacprzyk, J., Krawczak, M., Szmidt, E., Wygralak, M., Zadrożny, S. (eds.): Recent Advances in Fuzzy Sets, Intuitionistic Fuzzy Sets, Generalized Nets and Related Topics. Foundations, vol. I. SRI Polish Academy of Sciences, Warsaw (2011)

19. Atanassov, K., Baczyński, M., Drewniak, J., Kacprzyk, J., Krawczak, M., Szmidt, E., Wygralak, M., Zadrożny, S. (eds.): New Developments in Fuzzy Sets, Intuitionistic Fuzzy Sets, Generalized Nets and Related Topics. Foundations, vol. I. SRI Polish Academy of Sciences, Warsaw (2012)
20. Atanassov, K., Christov, R.: New conservative extensions of the generalized nets. Advances in Modelling and Analysis 14(2), 27–34 (1993)
21. Atanassov, K., Gluhchev, G., Hadjitodorov, S., Kacprzyk, J., Shannon, A., Szmidt, E., Vassilev, V.: Generalized Nets Decision Making and Pattern Recognition. Warsaw School of Information Technology Press, Warsaw (2006)
22. Atanassov, K., Homenda, W., Hryniewicz, O., Kacprzyk, J., Krawczak, M., Nahorski, Z., Szmidt, E., Zadrożny, S. (eds.): Developments in fuzzy sets, intuitionistic fuzzy sets, generalized nets and related topics. Applications, vol. II. SRI Polish Academy of Sciences, Warsaw (2010)
23. Atanassov, K., Homenda, W., Hryniewicz, O., Kacprzyk, J., Krawczak, M., Nahorski, Z., Szmidt, E., Zadrożny, S. (eds.): Recent Advances in Fuzzy Sets, Intuitionistic Fuzzy Sets, Generalized Nets and Related Topics. Applications, vol. II. SRI Polish Academy of Sciences, Warsaw (2011)
24. Atanassov, K., Homenda, W., Hryniewicz, O., Kacprzyk, J., Krawczak, M., Nahorski, Z., Szmidt, E., Zadrożny, S. (eds.): New Developments in Fuzzy Sets, Intuitionistic Fuzzy Sets, Generalized Nets and Related Topics. Applications, vol. II. SRI Polish Academy of Sciences, Warsaw (2012)
25. Atanassov, K., Hryniewicz, O., Kacprzyk, J., Krawczak, M., Nahorski, Z., Szmidt, E., Zadrożny, S. (eds.): Advances in Fuzzy Sets, Intuitionistic Fuzzy sets, Generalized Nets and Related Topics. Foundations, vol. I. EXIT Academic Press, Warsaw (2008)
26. Atanassov, K., Hryniewicz, O., Kacprzyk, J., Krawczak, M., Nahorski, Z., Szmidt, E., Zadrożny, S. (eds.): Advances in Fuzzy Sets, Intuitionistic Fuzzy sets, Generalized Nets and Related Topics. Applications, vol. II. EXIT Academic Press, Warsaw (2009)
27. Atanassov, K., Kacprzyk, J., Krawczak, M., Szmidt, E. (eds.): Issues in the representation and processing of uncertain and imprecise information. Fuzzy sets, intuitionistic fuzzy sets, generalized nets, and related topics. EXIT Academic Press, Warsaw (2005)
28. Atanassov, K., Kacprzyk, J., Krawczak, M., Szmidt, E. (eds.): Issues in intuitionistic fuzzy sets and generalized nets, vol. 3. Warsaw School of Information Technology Press, Warsaw (2006)
29. Atanassov, K., Kacprzyk, J., Krawczak, M., Szmidt, E. (eds.): Issues in intuitionistic fuzzy sets and generalized nets, vol. 4. Warsaw School of Information Technology Press, Warsaw (2007a)
30. Atanassov, K., Kacprzyk, J., Krawczak, M., Szmidt, E. (eds.): Issues in intuitionistic fuzzy sets and generalized nets, vol. 5. Warsaw School of Information Technology Press, Warsaw (2007b)
31. Atanassov, K., Kacprzyk, J., Krawczak, M., Szmidt, E. (eds.): Issues in intuitionistic fuzzy sets and generalized nets, vol. 6. Warsaw School of Information Technology Press, Warsaw (2008a)
32. Atanassov, K., Kacprzyk, J., Krawczak, M., Szmidt, E. (eds.): Issues in intuitionistic fuzzy sets and generalized nets, vol. 7. Warsaw School of Information Technology Press, Warsaw (2008b)

33. Atanassov, K., Kacprzyk, J., Krawczak, M., Szmidt, E. (eds.): Issues in intuitionistic fuzzy sets and generalized nets, vol. 8. Warsaw School of Information Technology Press, Warsaw (2010a)
34. Atanassov, K., Kacprzyk, J., Krawczak, M., Szmidt, E. (eds.): Issues in intuitionistic fuzzy sets and generalized nets, vol. 9. Warsaw School of Information Technology Press, Warsaw (2010b)
35. Atanassov, K., Kacprzyk, J., Krawczak, M., Szmidt, E. (eds.): Issues in intuitionistic fuzzy sets and generalized nets, vol. 10. Warsaw School of Information Technology Press, Warsaw (2011)
36. Atanassov, K., Krawczak, M.: Generalized nets modeling: A general concept. In: IFSR 2005 The New Roles of Systems Sciences for a Knowledge-Based Society, Kobe (November 2005)
37. Atanassov, K., Sotirov, S.: Optimization of a neural network of self-organizing maps type with time-limits by a generalized net. Advanced Studies in Contemporary Mathematics 13(2), 213–220 (2006)
38. Atanassov, K., Sotirov, S., Antonov, A.: Generalized Net Model for Parallel Optimization of Feed-forward Neural Network. Advanced Studies in Contemporary Mathematics 15(1), 109–119 (2007)
39. Atanassov, K., Sotirov, S., Krawczak, M.: Generalized net model of the intuitionistic fuzzy feedforward neural networks. Notes on Intuitionistic Fuzzy Sets 15(2), 18–23 (2009)
40. Avila, J.H.: The Feasibility of Continuation Methods for Nonlinear Equations. SIAM J. Numerical Analysis 11(1) (1974)
41. Azoff, E.M.: Neural Network Time Series Forecasting of Financial Markets. John Wiley and Sons, Chichester (1994)
42. Baldi, P., Hornik, K.: Neural networks and principal component analysis: Learning from examples without local minima. Neural Networks 2, 53–58 (1989)
43. Baum, E., Haussler, D.: What size net gives valid generalization. Neural Computation 1(1), 151–160 (1989)
44. Bellman, R.: Dynamic Programming. Princeton Univ. Press, Princeton (1972)
45. Bellman, R., Dreyfus, S.: Applied Dynamic Programming. Princeton Univ. Press, Princeton (1962)
46. Bertsekas, D.P.: Dynamic Programming and Optimal Control. Athena Scientific, Belmont (1995)
47. Bertsekas, D.P., Tsitsiklis, J.N.: Neuro-Dynamic Programming. Athena Scientific, Belmont (1996)
48. Bichta, M.: Computer implementation of some new neural networks learning algorithms. B. D. Dissertation. Warsaw School of Information Technology, Warsaw (2002)
49. Bichta, M.: Computer implementation of a new neural networks learning algorithms based on DDP. M.Sc Dissertation. Warsaw School of Information Technology, Warsaw (2006)
50. Brooks, S.P., Morgan, J.T.: Optimization using simulated annealing. The Statistician 44(2), 241–267 (1995)
51. Bryson, A.E., Ho, Y.-C.: Applied Optimal Control. Blasdel, Waltham (1969)
52. Campolucci, P., Marchegiani, A., Uncini, A., Piazza, F.: Signal-flow-graph derivation of on-line gradient learning algorithms. In: Proc. ICNN 1997, Houston, pp. 1884–1889 (June 1997)

53. Chernousko, F.L., Lyubushin, A.A.: Method of successive approximations for solution of optimal control problems. Optimal Control Applications and Methods 3, 101–114 (1982)
54. Chountas, P., Kolev, B., Rogova, E., Tasseva, V., Atanassov, K.: Generalized nets in artificial intelligence. Generalized Nets, Uncertain Data and Knowledge Engineering, vol. 4. Prof. M. Drinov Academic Publishing House, Sofia (2007)
55. Choy, E., Krawczak, M., Shannon, A., Szmidt, E. (eds.): A Survey of Generalized Nets. Raffles KvB Monograph, Sydney (2007)
56. Christofides, N.: Graph Theory. Academic Press, New York (1975)
57. Chua, L.O., Lin, P.-M.: Computer-Aided Analysis of Electric Circuits. Prentice-Hall, New Jersey (1975)
58. Crespi-Reghizzi, S., Mandrioli, D.: Some algebraic properties of Petri nets. Alma Erequeza XLV(2) (1976)
59. Cruz, J.B.: Feedback Systems. PWN Warszawa (1977) (in Polish)
60. Cybenko, G.: Approximation by superpositions of a sigmoidal function. Mathematics of Control, Signals, and Systems 2, 303–316 (1989)
61. Davidenko, D.: On the Approximate Solution of a System of Nonlinear Equations. Doklady Akad. SSSR 88, 601–604 (1953) (in Russian)
62. Dimitrova, S., Dimitrova, L., Kolarova, T., Petkov, P., Atanassov, K., Christov, R.: Generalized net models of the activity of NEFTOCHIM Petrochemical Combine in Bourgas. In: Atanassov, K. (ed.) Applications of Generalized Nets, pp. 208–213. World Scientific Publ. Co., Singapore (1993)
63. Dincheva, E., Atanassov, K.: Operator aspect of the theory of generalized nets. II. AMSE Review 12(4), 59–64 (1990)
64. Dreyfus, S.E.: Artificial neural networks, backpropagation, and the Kelley-Bryson gradient procedure. Journal of Guidance 13(5), 926–928 (1990)
65. Duch, W., Korbicz, J., Rutkowski, L., Tadeuisiewicz, R. (eds.): Neural Networks. Polish Academy of Sciences. EXIT Press, Warsaw (1999) (in Polish)
66. Dyer, P., McReynolds, S.R.: The Computation and Theory of Optimal Control. Academic Press, New York (1970)
67. Ellacott, S.W.: The numerical analysis approach. In: Taylor, J.G. (ed.) Mathematical Approaches to Neural Network, pp. 103–137. Elsevier Science Publishers B. V. (1993)
68. Ellacott, S.W.: Techniques for the mathematical analysis of neural networks. Journal of Computational and Applied Mathematics 50, 283–297 (1994)
69. Etzion, T., Yoeli, M.: Super nets and their hierarchy. Theory Comp. Sci. 23, 243–272 (1983)
70. Fettweiss, A.: A general theorem for signal flow networks, with applications. Digital Signal Processing, 126–130 (1972)
71. Findeisen, W., Szymanowski, J., Wierzbicki, A.: Computational Methods of Optimisation. WPW Warszawa (1972) (in Polish)
72. Fine, T.L.: Feedforward Neural Network Methodology. Springer, New York (1999)
73. Funahashi, K.: On the approximate realization of continuous mappings by neural networks. Neural Networks 2(3), 183–192 (1989)
74. Genrich, H., Lautenbach, K.: The analysis of distributed systems by means of predicate/transition nets. In: Kahn, G. (ed.) Semantics of Concurrent Computation. LNCS, vol. 70, pp. 123–146. Springer, Heidelberg (1979)

75. Geoffrion, A.M.: StrictlycConcave parametric programming, Part II: Additional Theory and Computational Considerations. Management Science (Series A) 13(5), 359–370 (1967)
76. Hadjyisky, L., Atanassov, K.: Theorem for representation of the neu-ronal networks by generalized nets. AMSE Review 12(3), 47–54 (1990a)
77. Hadjyisky, L., Atanassov, K.: A generalized net, representing the elements of one neuron network set. AMSE Review 14(4), 55–59 (1990b)
78. Hadjyisky, L., Atanassov, K.: Generalized nets representing the ele-ments of neuron networks. In: Atanassov, K. (ed.) Applications of Generalized Nets, pp. 49–67. World Scientific Publ. Co., Singapore (1993)
79. Hadjyisky, L., Atanassov, K.: Generalized net model of the intuitionistic fuzzy neural networks. Advances in Modelling and Analysis 23(2), 59–64 (1995)
80. Hassoun, M.H.: Fundamentals of Artificial Neural Networks. MIT Press (1995)
81. Hebb, D.O.: The Organization of Behaviour. John Wiley and Sons, New York (1949)
82. Hecht-Nielsen, R.: Neurocomputing. Addison-Wesley, Massachusetts (1990)
83. Hecht-Nielsen, R.: Theory of the backpropagation network. In: Wechsler, H. (ed.) Neural Networks for Perception. Academic Press, New York (1992)
84. Hertz, J.A., Krogh, A., Palmer, R.G.: The Theory of Neural Computation. Addison-Wesley, Reading (1990)
85. Hinton, G.E., Sejnowski, T.J.: Optimal Perceptual Inference. In: Proc. of IEEE Conference on Computer Vision and Pattern Recognition, Washington, pp. 448–453 (1983)
86. Hopfield, J.J.: Neural networks and physical systems with emergent collective computational properties. Proc. National Academy of Science USA 79, 2554–2558 (1982)
87. Hopfield, J.J., Tank, D.: Neural computation of decisions in optimisation problems. Biological Cybernetics 52, 141–152 (1985)
88. Hornik, K.: Neural networks and principal component analysis: Learning from examples without local minima. Neural Networks 2, 53–58 (1989)
89. Hornik, K.: Some new results on neural network approximation. Neural Networks 6(8), 1069–1072 (1993)
90. Hornik, K., Stinchcombe, M., White, H.: Multilayer feedforward networks are universal approximators. Neural Networks 2, 359–366 (1989)
91. http://ifigenia.org/wiki/
92. Irie, B., Miyake, S.: Capabilities of three-layer perceptrons. In: Proc. IEEE International Conference on Neural Networks, San Diego, vol. 1, pp. 641–648 (1988)
93. Ito, Y.: Representation of functions by superpositions of a step or sigmoid function and their applications to neural network theory. Neural Networks 4(3), 385–394 (1991)
94. Jacobs, R.A.: Increased rates of convergence through learning rate adaptation. Neural Networks 1, 295–307 (1998)
95. Jacobson, D.H., Mayne, D.Q.: Differential Dynamic Programming. Am. Elsevier Pub. Comp., New York (1970)
96. Jensen, K.: Coloured Petri nets and the invariant method. Theoretical Computer Science 14(3), 317–336 (1981)
97. Kacprzyk, J., Krawczak, M., Atanassov, K. (eds.): Issues in intuitionistic fuzzy sets and generalized nets, vol. 2. Warsaw School of Information Technology Press, Warsaw (2005)

98. Katsuura, H., Sprecher, D.A.: Computational aspects of Kolmogorov's superposition theorem. Neural Networks 7(3), 455–461 (1994)
99. Kelley, H.J.: Gradient theory of optimal flight paths. ARS Journal 30(10), 947–954 (1960)
100. Kirkpatrick, S., Gelatt, C.D., Vecchi, M.P.: Optimisation by simulated annealing. Science 220, 671–680 (1983)
101. Kirpkpatrick, S.: Optimization by simulated annealing: Quantitative studies. J. Statist. Physics 34, 975–986 (1984)
102. Kohonen, T.: Self-organized formation of topologically correct feature maps. Biological Cybernetics 43, 59–69 (1982)
103. Kohonen, T.: Self-organizing and Associative memory. Springer, Berlin (1987)
104. Kohonen, T.: Self-organizing maps. Springer, Berlin (1997)
105. Kolev, B., El-Darzi, E., Sotirova, E., Petronias, I., Atanassov, K., Chountas, P., Kodogiannis, V.: Generalized nets in artificial intelligence. Generalized Nets, Relational Data Bases and Expert Systems, vol. 3. Prof. M. Drinov Academic Publishing House, Sofia (2006)
106. Kolmogorov, A.N.: On the representation of continuous functions of many variables by superposition of continuous functions of one variable and addition. Doklady Akad. Nauk USSR 114, 953–956 (1957)
107. Koneva, L., Atanassov, K.: Generalized net with optimisation components. AMSE Review 13(1), 25–30 (1990a)
108. Koneva, L., Atanassov, K.: Generalized net model of transportation problem. AMSE Review 13(1), 17–23 (1990b)
109. Korbicz, J., Obuchowicz, A., Uciński, D.: Artificial Neural Networks. PLJ, Warszawa (1994) (in Polish)
110. Kotov, V.: An algebra for parallelism based on Petri nets. In: Winkowski, J. (ed.) MFCS 1978. LNCS, vol. 64, pp. 39–55. Springer, Heidelberg (1978)
111. Krawczak, M.: Differential dynamic programming of first order as backpropagation interpretation. In: Proc. Int. Conference on Systems Science, Wrocław (September 1995a)
112. Krawczak, M.: Iterative minimax method for supervised multilayer neural networks. In: Proc. Int. Conference on Methods and Models in Automation and Robotics. IEEE, Międzyzdroje (1995b)
113. Krawczak, M.: Neural networks learning as a multiobjective optimal control problem. Mathware and Soft Computing 4(3), 195–202 (1997)
114. Krawczak, M.: Game approach to neural networks learning. In: Proc. 8 Int. Symposium on Dynamic Games and Applications (IEEE, IFAC, INRIA), Maastricht (July 1998)
115. Krawczak, M.: Dynamic learning for feedforward neural networks. In: Proc. ICONIP 1999 ANZIIS 1999 & ANNES 1999 & ACNN 1999, Perth, Australia, pp. 1057–1062 (November 1999a)
116. Krawczak, M.: Dynamic programming and fuzzy reinforcement of backpropagation for interest rate prediction. In: Ribeiro, R.A., Zimmerman, H.J., Yager, R.R., Kacprzyk, J. (eds.) Soft Computing in Financial Engineering, pp. 142–155. Springer (1999b)
117. Krawczak, M.: Backpropagation versus dynamic programming approach. Bulletin of Polish Academy of Sciences 48(2), 167–180 (2000a)
118. Krawczak, M.: Feedforward neural networks learning by continua-tion method. In: Proc. IIZUKA 2000, Fukuoka (October 2000b)

119. Krawczak, M.: Neural networks learning and homotopy method. In: Proc. 5 Int. Conference on Neural Networks and Soft Computing, pp. 94–99. IEEE, Zakopane (2000c)
120. Krawczak, M.: Neural networks learning by continuation method. In: Proc. 6 Int. Conference on Methods and Models in Automation and Robotics, pp. 727–732. IEEE, Międzyzdroje (2000d)
121. Krawczak, M.: Neural networks learning by homotopy method. In: Proc. ICONIP 2000, Taejin, Korea (October 2000e)
122. Krawczak, M.: Neural networks learning as a particular optimal control problem. In: Proc. 7 Int. Conference on Methods and Models in Automation and Robotics. IEEE, Międzyzdroje (2001a)
123. Krawczak, M.: Parameterisation of Neural Networks Learning. In: Proc. 14 Int. Conference on Systems Science, Wrocław, Poland, pp. 330–337 (September 2001b)
124. Krawczak, M.: Adjoint multilayer neural networks. Gutenbaum, J. (ed.) Automatic Control and Management, SRI Polish Academy of Sciences, Special Issue, Warsaw, pp. 153–166 (2002a)
125. Krawczak, M.: Heuristic dynamic programming for neural networks learning, Part 1: Learning as a Control Problem. Advances in Soft Computing, pp. 218–223. Springer (2002b)
126. Krawczak, M.: Heuristic dynamic programming for neural networks learning, Part 2: I-order Differential dynamic programming. Advances in Soft Computing, pp. 224–229. Springer (2002c)
127. Krawczak, M.: Boltzman machine built by generalized nets. In: Cha, J., Jardim-Goncalves, R., Steiner-Garcao, A. (eds.) Concurrent Engineering, pp. 1095–1099. Balkema Publishers, Tokyo (2003a)
128. Krawczak, M.: Generalized nets representation of multilayer neural networks. Special Testimony Commission on Electronics and Computer Science, Sofia (2003b)
129. Krawczak, M.: Generalized nets representation of feedforward neural networks simulation process. Advanced Studies on Contemporary Mathematics 7(1), 69–86 (2003c)
130. Krawczak, M.: Modelling of adjoint neural networks by generalized nets. In: Proc. 9th IEEE International Conference on Methods and Models in Automation and Robotics MMAR 2003, Międzyzdroje, Poland, pp. 33–42. Szczecin Technical University Press (August 2003d)
131. Krawczak, M.: Multilayer Neural Systems and Generalized Net Models. EXIT Academic Press, Warsaw (2003e)
132. Krawczak, M.: An example of generalized nets application to modelling of neural networks simulation. In: De Baets, B., De Caluwe, R., De Tre, G., Fodor, J., Kacprzyk, J., Zadrożny, S. (eds.) Current Issues in Data and Knowledge Engineering, pp. 297–308. EXIT Academic Press, Warsaw (2004a)
133. Krawczak, M.: Generalized net modelling concept-neural networks models. In: Studziński, J., Drelichowski, L., Hryniewicz, O. (eds.) Computer aid Economy and Enviromental Development, pp. 203–216. SRI Polish Academy of Sciences, Warsaw (2004b)
134. Krawczak, M.: On a new way to model discrete dynamic systems. In: Kulikowski, R., Kacprzyk, J., Słowiński, R. (eds.) Operation and System Research. Decision Making, pp. 401–412. EXIT Academic Press, Warsaw (2004c)
135. Krawczak, M.: Neural networks learning as a parameterized optimal control problem. In: Grzegorzewski, P., Krawczak, M., Zadrożny, S. (eds.) Soft Computing. Tools, Techniques and Applications, pp. 145–156. EXIT Academic Press, Warsaw (2004d)

136. Krawczak, M.: On a different way to get gradient-based neural network learning algorithms. In: Proc. 10th IEEE International Conference on Methods and Models in Automation and Robotics MMAR 2004, Miedzyzdroje, pp. 1297–1302. Szczecin Technical Univ. Press (2004e)
137. Krawczak, M.: On a generalized net model of MLNN simulation. In: Grzegorzewski, P., Krawczak, M., Zadrożny, S. (eds.) Soft Computing. Tools, Techniques and Applications, pp. 157–172. EXIT Academic Press, Warsaw (2004f)
138. Krawczak, M.: On a way to gradient-based neural networks learning algorithms. In: Studziński, J., Drelichowski, L., Hryniewicz, O. (eds.) Computer Aiding of Social, Economical and Environment Development, Part 2, pp. 23–35. SRI Polish Academy of Sciences, Warsaw (2004g)
139. Krawczak, M.: A way to aggregate multilayer neural networks. In: Duch, W., Kacprzyk, J., Oja, E., Zadrożny, S. (eds.) ICANN 2005, Part II. LNCS, vol. 3697, pp. 19–24. Springer, Heidelberg (2005a)
140. Krawczak, M.: Generalized net models of MLNN learning algorithms. In: Duch, W., Kacprzyk, J., Oja, E., Zadrożny, S. (eds.) ICANN 2005, Part II. LNCS, vol. 3697, pp. 25–30. Springer, Heidelberg (2005b)
141. Krawczak, M.: Generalized nets conception of modeling: MLNN models. In: Proc. 11 IEEE International Conference on Methods and Models in Automation and Robotics MMAR 2005, Miedzyzdroje, pp. 787–793. Szczecin Technical Univ. Press (August 2005c)
142. Krawczak, M.: Generalized nets representation of multilayer neural networks simulation process. In: Kacprzyk, J., Nahorski, Z., Wagner, D. (eds.) Systems Research Applications in Science, Technology and Economy, pp. 279–294. EXIT Academic Press, Warsaw (2005d)
143. Krawczak, M.: Modelling of adjoint neural networks by generalized net. In: Atanassov, K., Kacprzyk, J., Krawczak, M., Szmidt, E. (eds.) Issues in the Representation and Processing of Uncertain and Imprecise Information. Fuzzy Sets, Intuitionistic Fuzzy Sets, Generalized Nets, and Related Topics, pp. 217–227. EXIT Academic Press, Warsaw (2005e)
144. Krawczak, M.: A novel modeling methodology: generalized nets. In: Cader, A., Rutkowski, L., Tadeusiewicz, R., Żurada, J. (eds.) Artificial Intelligence and Soft Computing, pp. 1160–1168. Publishing House EXIT, Warsaw (2006a)
145. Krawczak, M.: Algebraic aspects of generalized nets. In: Studziński, J., Drelichowski, L., Hryniewicz, O. (eds.) Development of Methods and Technologies of Informatics for Process Modeling and Management, pp. 37–47. SRI Polish Academy of Sciences, Warsaw (2006b)
146. Krawczak, M., Aladjov, H.: Generalized net model of backpropagation algorithm. In: Proc. of 3 International Workshop on Generalized Nets, Sofia, pp. 32–36 (October 2002)
147. Krawczak, M., Aladjov, H.: Generalized net model of adjoint neural networks. Advanced Studies on Contemporary Mathematics 7(1), 20–32 (2003)
148. Krawczak, M., Atanassov, K., Sotirov, S.: Generalized net model for parallel optimization of feed-forward neural network with variable learning rate backpropagation algorithm with time limit. In: Sgurev, V., Hadjiski, M., Kacprzyk, J. (eds.) Intelligent Systems: From Theory to Practice. SCI, vol. 299, pp. 361–371. Springer, Heidelberg (2010)

149. Krawczak, M., El-Darzi, E., Atanassov, K., Tasseva, V.: Generalized net for control and optimization of real processes through neural networks using intuitionistic fuzzy estimations. Notes on Intuitionistic Fuzzy Sets 12(2), 54–60 (2007)
150. Krawczak, M., Gocheva, E., Sotirov, S.: Modelling of the verification by iris scanning by generalized net. In: Atanassov, K., Hryniewicz, O., Kacprzyk, J., Krawczak, M., Nahorski, Z., Szmidt, E., Zadrożny, S. (eds.) Advances in Fuzzy Sets, Intuitionistic Fuzzy Sets, Generalized Nets and Related Topics, pp. 69–74. EXIT Academic Press, Warsawa (2008)
151. Krawczak, M., Mizukami, K.: The control theory approach to perceptron learning process. IEE of Japan, Okayama (1994)
152. Krawczak, M., Sotirov, S., Atanassov, K.: Multilayer Neural Networks and Generalized Nets. Warsaw School of Information Technology Press, Warsaw (2010)
153. Krawczak, M., Sotirov, S., Sotirova, E.: Generalized net model for parallel optimization of MLNN with momentum BP algorithm with time Limit. In: Proc. IEEE 6th International Conference on Intelligent Systems 2012, Sofia, pp. 233–236 (2012)
154. Kuncheva, L., Atanassov, K.: An intuitionistic fuzzy RBF network. In: Proceedings of EUFIT 1996, Aachen, pp. 777–781 (September 1996)
155. Kurkova, V.: Kolmogorov's theorem and multilayer neural networks. Neural Networks 5(3), 501–506 (1992)
156. Larson, R.E., Korsak, A.J.: A dynamic programming successive approximations technique with convergence proofs, Part 1: Description of the method and application. Automatica 6, 245–252 (1970)
157. Lawrence, S., Giles, C.L., Tsoi, A.C.: What size neural network gives optimal generalization? Convergence properties of backpropagation. Technical Report, Univ. of Maryland, UMIACS-TR-96-22, CS-TR-3617 (1996)
158. Li, D., Haimes, Y.Y.: Multilevel methodology for a class of nonseparable optimisation problems. International Journal of Systems Science 21, 2352–2360 (1990)
159. Li, D., Haimes, Y.Y.: Extension of dynamic programming to nonseparable problems. Computer and Mathematics with Applications 21, 51–56 (1991)
160. Light, W.A.: Ridge functions, sigmoidal functions and neural networks. In: Approximation Theory VII, pp. 163–206. Academic Press, Boston (1992)
161. Luenberger, D.G.: Linear and Nonlinear Programming. Addison-Wesley, Massachusetts (1984)
162. Masters, T.: Practical Neural Network Recipes in C++. Academic Press, San Diego (1993)
163. Mayne, D.: A second order gradient method for determining optimal trajectories of non-linear discrete-time systems. International J. Control 3, 85–95 (1966)
164. McCulloch, W.S., Pitts, W.: A logical calculus of the ideas immanent in nervous activity. Bulletin of Mathematical Biophisics 9, 127–147 (1943)
165. Melo-Pinto, P., Kim, T., Atanassov, K., Sotirova, E., Shannon, A., Krawczak, M.: Generalized net model of e-learning evaluation with intuitionistic fuzzy estimations. In: Atanassov, K., Kacprzyk, J., Krawczak, M., Szmidt, E. (eds.) Issues in the Representation and Processing of Uncertain Imprecise Information: Fuzzy Sets, Intuitionistic Fuzzy Sets, Generalized Nets, and Related Topics, pp. 241–249. EXIT Academic Press, Warsaw (2005)
166. Mesarovic, M., Takahara, Y.: General System Theory: Mathematical Foundation. Academic Press, New York (1975)
167. Mihalewicz, Z.: Genetic Algorithms + Data Structures = Evolution Programs. Springer, Heidelberg (1999)

168. Minsky, M., Papert, S.: Perceptrons. MIT Press, Cambridge (1969)
169. Morrison, F.: Dynamic Systems Modelling. PWN, Warsaw (1996) (in Polish)
170. von Neumann, J.: Theory of Self-Reproducing Automata. Machine. Urbana Univ. Press (1966)
171. Nikolov, N.: Generalized Nets and Semantic Networks. Advances in Modelling and Analysis 27(1), 19–25 (1995)
172. Ohno, K.: A new approach to differential dynamic programming for discrete time systems. IEEE TAC 23(1), 37–47 (1978)
173. Ortega, J.M., Rheinboldt, W.C.: Iterative Solution of Nonlinear Equations in Several Variables. Academic Press, N.Y. (1970)
174. Osowski, S.: Neural Networks for Information Processing. WPW, Warszawa (2000) (in Polish)
175. Pawlak, Z.: Personal communication (2005)
176. Pedrycz, W.: Computational Intelligence. CRC Press, N.Y. (1998)
177. Penfield, P., Spence, R., Duiker, S.: Tellegen's Theorem and Electrical Networks. MIT Press, Cambridge (1970)
178. Petri, A.C.: Kommunication mit Automaten, Ph.D. Diss., Univ. of Bonn (1962); Schriften des Inst. fur Instrument. Math. (2) (1962)
179. Petri, A.C.: Net Theory and Applications. LNCS, vol. 84, pp. 1–20 (1980)
180. Pitts, W., McCulloch, W.S.: How we know universals: the perception of auditory and visual forms. Bulletin of Mathematical Biophysics 9, 127–147 (1947)
181. Polak, E.: Computational Methods in Optimisation. Academic Press, New York (1971)
182. Radeva, V., Krawczak, M., Choy, E.: Review and bibliography on generalized nets theory and applications. Advanced Studies in Contemporary Mathematics 4(2), 173–199 (2002)
183. Rashevsky, N.: Mathematical Biophysics. University of Chicago Press, Chicago (1948)
184. Richter, S., de Carlo, R.: A homotopy method for eigenvalue assignment using decentralized state feedback. IEEE Transactions on Automatic Control AC-29(2) (1984)
185. Roitenberg, J.N.: Control Theory. PWN Warszawa (1978) (in Polish)
186. Rosenblat, F.: The perceptron: A probalistic model for information storage and organization in the brain. Psychological Review 65, 386–408 (1958)
187. Rumelhart, D.E., Hilton, G.E., Williams, R.J.: Learning internal representations by error propagation. In: Rumelhart, D.E., Williams, R.J. (eds.) Parallel Distributed Processing: Explorations in the Microstructure of Cognition, pp. 318–362. MIT Press, Cambridge (1986)
188. Saratchandran, P.: Dynamic programming approach to optimal weight selection in multilayer neural networks. IEEE Trans. Neural Networks 2(4), 465–467 (1991)
189. Schoen, F.: Stochastic techniques for global optimisation: A survey of recent advances. Journal of Global Optimisation 1, 207–228 (1991)
190. Sethi, P.S., Thompson, G.L.: Optimal Control Theory: Applications to Management Science and Economics. Kluwer Academic Publishers (1981)
191. Shannon, A., Atanassov, K., Orozova, D., Krawczak, M., Sotirova, E., MeloPinto, P., Petrounias, I., Kim, T.: Generalized net s and information flow within a university. Warsaw School of Information Technology Press, Warsaw (2007)

192. Shannon, A., Langova-Orozova, D., Sotirova, E., Petrounias, I., Atanassov, K., Krawczak, M., Melo-Pinto, P., Kim, T.: Generalized Net Modelling of University Processes. KvB Monograph, Sydney (2005)
193. Shannon, A., Roeva, O., Pencheva, T., Atanassov, K.: Generalized Nets Modelling of Biotechnological Processes. "Prof. M. Drinov" Academic Publishing House, Sofia (2004)
194. Shapiro, S.: A stochastic Petri net with applications to modelling occupancy timed for concurrent task systems. Networks 9, 375–379 (1979)
195. Smagt van der, P.P.: Minimisation methods for training feedforward neural networks. Neural Networks 7(1), 1–11 (1994)
196. Sotirov, S.: Modeling the algorithm backpropagation for training of neural networks with generalized nets. Part 1. In: Proc. 4th International Workshop on Generalized Nets, Sofia, September 23, pp. 61–67 (2003)
197. Sotirov, S.: A method of accelerating neural network training. Neural Processing Letters 22(2), 163–169 (2005)
198. Sotirov, S.: Modeling the accelerating backpropagation algorithm with generalized nets. Advanced Studies in Contemporary Mathematics 9(2), 217–225 (2006)
199. Sotirov, S., Krawczak, M.: Modelling the work of self organizing neural network with generalized nets. In: Issues in Intuitionistic Fuzzy Sets and Generalized Nets, vol. 3, pp. 57–64. Warsaw School of Information Technology Press, Warsaw (2003)
200. Sotirov, S., Krawczak, M.: Modeling the algorithm backpropagation for learning of neural networks with generalized networks, Part 2. In: Issues in Intuitionistic Fuzzy Sets and Generalized Nets, vol. 3, pp. 65–69. Warsaw School of Information Technology Press, Warsaw (2006)
201. Sotirov, S., Krawczak, M.: Generalized net model of the art neural networks, Part 2. In: Issues in Intuitionistic Fuzzy Sets and Generalized Nets, vol. 7, pp. 67–74. Warsaw School of Information Technology Press, Warsaw (2008a)
202. Sotirov, S., Krawczak, M.: Generalized net model of the art neural networks, Part 2. In: Issues in Intuitionistic Fuzzy Sets and Generalized Nets, vol. 7, pp. 75–82. Warsaw School of Information Technology Press, Warsaw (2008b)
203. Sotirov, S., Krawczak, M.: Modelling a layered digital dynamic network by a generalized net. In: Issues in Intuitionistic Fuzzy Sets and Generalized Nets, pp. 84–91. Warsaw School of Information Technology Press, Warsaw (2011)
204. Sotirov, S., Krawczak, M.: Generalized net model for parallel optimization of multilayer perceptron with conjugate gradient backpropagation algorithm. In: Atanassov, K., Baczyński, M., Drewniak, J., Kacprzyk, J., Krawczak, M., Szmidt, E., Wygralak, M., Zadrożny, S. (eds.) New Developments in Fuzzy Sets, Intuitionistic Fuzzy Sets, Generalized Nets and Related Topics. Foundations, vol. I, pp. 179–190. SRI Polish Academy of Sciences (2012)
205. Sotirov, S., Krawczak, M., Kodogiannis, V.: Modeling the work of learning vector quantization neural networks. In: Proc. 7 Int. Workshop on Generalized Nets, Sofia, pp. 39–44 (July 2006)
206. Sotirov, S., Krawczak, M., Kodogiannis, V.: Generalized nets model of the Grossberg neural networks, Part 1. In: Issues in Intuitionistic Fuzzy Sets and Generalized Nets, vol. 4, pp. 27–34. Warsaw School of Information Technology Press, Warsaw (2007)
207. Sotirova, E.: Generalized nets as tools for modelling of open, hybrid and closed systems: an example with expert system. Studies on Contemporary Mathematics 13(2), 221–234 (2006)

208. Starke, P.: Petri-Netze. VEB Deutscher Verlag der Wissenschaften, Berlin (1980)
209. Sun, X., Cheney, E.W.: The fundamentals of sets of ridge functions. Aequationes Math. 44, 226–235 (1992)
210. Tadeusiewicz, R.: Neural Networks. RM, Warsaw (1993) (in Polish)
211. Tang, Z., Koehler, G.J.: Deterministic global optimal FNN raining algorithms. Neural Networks 7(2), 301–311 (1994)
212. Wan, E., Beaufays, F.: Diagrammatic derivation of gradient algorithms for neural networks. Neural Computation 8(1), 182–206 (1996)
213. Werbos, P.J.: Maximizing long-term gas industry profits in two minutes in Lotus using neural network methods. IEEE Trans. on Systems, Man, and Cybernetics 19(2), 315–333 (1989)
214. Widrow, B., Hoff, M.: Adaptive Switching Circuits. In: 1960 IRE WESCON Convention Record, pp. 96–104. IRE, New York (1960)
215. Yakowitz, S., Rutherford, B.: Computational aspects of discrete-time optimal control. Applied Mathematics and Computation 15, 29–45 (1984)
216. Zerros, C., Irani, K.: Colored Petri nets: their properties and applications. Systems Engineering Lab. TR 107, Univ. of Michigan (1997)
217. Zurada, J.M.: Introduction to Artificial Neural Systems. West Publishing Company, St. Paul (1992)
218. Zurada, J.M.: Artificial Neural Networks. PWN, Warszawa (1996) (in Polish)

Printed in the United States
By Bookmasters